50 Years of DNA

Edited by
Julie Clayton and Carina Dennis

Foreword by
Philip Campbell

nature ✳ palgrave
macmillan

Acknowledgements

At the journal *Nature* I would like to thank Maxine Clarke and Christopher Surridge for their enthusiastic support and scientific input, Barbara Izdebska for her thoroughness and high standards in picture research, and Simon Gribben for his assistance with the reproduction of the previously published papers. I am also extremely grateful to Professor Sir Aaron Klug at the Laboratory for Molecular Biology, Cambridge, for in-depth discussion about Rosalind Franklin's X-ray diffraction studies, and to the staff of the Cold Spring Harbor Laboratory, Long Island, for their hospitality and assistance during the research of this book. Enormous thanks also go to the editorial production team that have kept the project on the rails: at Palgrave Macmillan, Josie Dixon; and at Book Production Consultants in Cambridge, project manager Sue Gray, copy editor Jenny Knight, production assistant Emma Forster and graphic designer Tim McPhee.

Julie Clayton

First published 2003 by
PALGRAVE MACMILLAN
Houndmills, Basingstoke, Hampshire RG21 6XS and
175 Fifth Avenue, New York, N.Y. 10010
Companies and representatives throughout the world

PALGRAVE MACMILLAN is the global academic imprint of the Palgrave Macmillan division of St. Martin's Press, LLC and of Palgrave Macmillan Ltd. Macmillan® is a registered trademark in the United States, United Kingdom and other countries. Palgrave is a registered trademark in the European Union and other countries.

ISBN 1-4039-1479-6 hardback
ISBN 1-4039-1480-X paperback

This book is printed on paper suitable for recycling and made from fully managed and sustained forest sources.

A catalogue record for this book is available from the British Library.

A catalog record for this book is available from the Library of Congress.

10 9 8 7 6 5 4 3 2 1
12 11 10 09 08 07 06 05 04 03

Produced in association with
Book Production Consultants plc, Cambridge, UK
Printed and bound in Italy

Contents

Nature essays

Foreword

It was one of the greatest moments in the history of science and of humanity, but not even the people involved knew for sure just how momentous it was, and the rest of the world was oblivious.

In the 25 April 1953 issue of *Nature*, without fanfare and almost wholly unremarked in the press at the time, appeared beautiful X-ray crystallographic measurements of the structure of deoxyribonucleic acid – DNA. As is the way with X-ray measurements, these indirect signatures required interpretation. Accompanying them was a model, which was only accepted subsequently as the model, that explained the observations by means of a proposed molecular structure: the double helix.

Now, looking back over five decades and peering forward as best we can, we can fully appreciate the scale of the revolution that subsequently unfolded. I want to express my appreciation here to Julie Clayton's fine and lively account of the historical moment itself and its consequences, and to the editorial dedication and skill of Carina Dennis in commissioning and editing the insightful celebratory articles that form the second half of this book.

As the historian Robert Olby describes on page 88, it took years for the double helix to be widely acknowledged, and there were other discoveries made at that time that were just as critical for biological understanding. But the sheer beauty of the DNA molecule, and the direct and easily visualized coupling between its structure and the essential mechanism of inheritance, go a long way to explain its iconic status and the reason why we give it such prominence in this book.

For those familiar only with the elegant double-helix structure extending like a spiral staircase, the DNA that sits in the trillions of cells in our bodies is for most of the time unrecognizable. As several authors in this book describe, biologists, physicists and chemists all have a long way to go in understanding the properties, dynamics and storage of the molecule and the way in which its roles in cells are regulated. Also unclear are the diverse ways in which the genetic makeup and the environment both influence the expression of the DNA's code – synthesizing molecules that in turn are the building blocks of us and of our more-or-less distant relatives in the kingdoms of life.

The first step in the uncovering of the biology of DNA could reasonably be identified as the discovery that the nuclei of certain cells contained an acid, then called 'nuclein'. Coincidentally, this occurred in the same year that *Nature* was launched, 1869. Since then *Nature* has published many of the key steps in DNA biology, including our 2001 publication of the draft sequence – the list of letters in the genetic alphabet strung out along the human genome's DNA – from the international consortium set up to achieve that task. As this book appears, we are continuing with publications of completed sequences of human chromosomes.

But given the way such biology has to be accomplished these days, these latest papers are all the outcome of insights in genetics and in sequencing methods and project plans made years ago. Perhaps more challenging and fun is to identify other papers appearing in *Nature* today that will come to be seen to be as fundamental to biology as were those papers on DNA published in 1953.

Philip Campbell
Editor, *Nature*

Preface

Fifty years since its discovery, the DNA double helix touches every area of science, medicine and culture. Its curving, twisting form is also an image that graces art, advertising, coins and stamps. It will have escaped the notice of few people in western society. This book celebrates its revelation, half a century ago, and includes a collection of exclusive interviews and essays by prominent experts in biology and medicine – all telling the extraordinary story of how the structure of life's most important molecule was discovered, and the impact this has today.

It begins with the tale of how two young and ambitious scientists, James Watson and Francis Crick, deftly pieced together the clues to make

"Science can always be used for harm. The question is, have we made our lives better over the past hundred years? I would say yes, and I would think that over the next hundred years we're going to make it better still."

James Watson, December 2002

their Nobel prize-winning discovery. Their paper announcing the double helix appeared in *Nature* on 25 April 1953, and is reproduced here, alongside those of the experimentalists whose work was critical to their insight: Rosalind Franklin and Maurice Wilkins. The personal tragedy that lies behind Rosalind Franklin's contribution is also described, together with a tribute to her remarkable scientific achievements.

The most important element of the Watson and Crick model was the specific pairing of nucleotides holding the double helix together. The discovery of this arrangement provided the first window onto how hereditary information is passed from generation to generation, and onto

what underlies susceptibility to disease. It gave rise to the tools with which we could discover our ancestry, solve murder cases, and understand human health and disease.

I have written an introductory series of chapters which take the reader on a 50-year journey through the explosion of molecular biology to the triumphant sequencing of the human genome, with testimonies by some of the major players, to illustrate the legacy of that first glimpse of base pairing. I describe how the key players of those early days have continued as luminaries and leaders in their respective fields, and what they hope genetics will achieve over the next fifty years.

As genome sequences tumble off the production lines of hundreds of labs worldwide, they are providing an ever-growing resource for biology and medicine – with the greatest impact yet to come, in disease diagnosis and therapy. In an interview, James Watson, who is no stranger to controversy, welcomes the future possibility of tinkering with our DNA not only to protect future generations from cancer, but also to alter human behaviour. *Nature* biology editor Christopher Surridge and science journalist Steve Nadis also describe how DNA has inspired whole new genres of fiction in books and film, as well as new material with which to create, literally, new works of art.

These chapters provide background and context for the essays (pages 82–139) planned by Philip Campbell (Editor in Chief of *Nature*) and Carina Dennis, of *Nature*, and developed and edited by Carina Dennis. These essays give expert views on the history and impact of Watson's and Crick's discovery, and guide readers towards an exciting future of which the double helix is just the beginning.

Julie Clayton, January 2003

On the shoulders of giants

Here's looking at you

We are obsessed with it – curly hair, straight hair, eye colour, the shapes of our noses and other body parts – our appearance is a constant source of fascination. New parents speculate endlessly about the resemblance between their babies and other family members. After that, their attention turns to behaviour, personality and aptitude: the temper tantrums of a toddler, the musical skills of a 7-year-old, the exam success of a 16-year-old. Who did those come from? Was it upbringing or was it inherited?

For centuries people hadn't a clue about how traits were passed on from generation to generation. Hippocrates (460–377 BC) had suggested the notion of 'pangenesis', in which each part of the parents' bodies somehow went into forming the equivalent body part of the child.

It was to be another two-and-a-half thousand years before the true nature of heredity was discovered.

The struggle for life

A five-year voyage around South America beginning in 1831 set the naturalist Charles Darwin on his way to shattering the comfortable Victorian notion that all plant and animal species had remained essentially unaltered since their day of creation. In 1859, after many years of thought and study, Darwin rushed out his publication *The Origin of Species by Means of Natural Selection*, to beat competition from another explorer and collector, Alfred Russel Wallace. Darwin's book outlined the stunning theory that every creature visible on Earth was the result of a 'struggle for existence'. He said that new species of animals and plants arose when competition favoured the 'survival of the fittest'. Variations between individuals enabled some to survive preferentially; by accident rather than design, some were better at securing a bigger meal, attracting a fitter mate or escaping a predator, and would pass their traits on to their offspring. Natural selection, rather than an act of creation, had produced all plants and animal varieties, including humans.

In *The Origin of Species*, Darwin wrote:

I could give many facts, showing how anxious bees are to save time; for instance, their habit of cutting holes and sucking the nectar at the bases of certain flowers, which they can, with a very little more trouble, enter by the mouth. Bearing such facts in mind, I can see no reason to doubt that an accidental deviation in the size and form of the body,

[Julie Clayton]

or in the curvature and length of the proboscis, far too slight to be appreciated by us, might profit a bee or other insect, so that an individual so characterised would be able to obtain its food more quickly, and so have a better chance of living and leaving descendants. Its descendants would probably inherit a tendency to a similar slight deviation of structure.

Knowing nothing of the substance involved (DNA), or the mechanism of variation (changes in the DNA alphabet called 'mutations'), Darwin did at least venture that 'the most frequent cause of variability may be attributed to the male and female reproductive elements having been affected prior to the act of conception'. He adapted Hippocrates' idea of pangenesis to suggest that elements from both parents were somehow blended together in the creation of offspring.

Blending out, genes in

Aware of Darwin's ideas, but studying in the relative obscurity of the Abbey of St Thomas Brno in Austria, the Augustinian monk Gregor

Below: Charles Darwin.

Right: Gregor Mendel (Abbey of St Thomas, Brno, Czech Republic)

"the most frequent cause of variability may be attributed to the male and female reproductive elements having been affected prior to the act of conception."

Charles Darwin

Mendel had already embarked on a course that would lead him to accept the theory of natural selection, but reject the proposed mechanism of pangenesis. Later dubbed 'father of genetics', Mendel experimented with garden peas, recording the appearance of seeds over many generations, such as whether they were wrinkled or round, yellow or green. In 1865 he presented a paper entitled 'Experiments in Plant Hybridization' to the Natural History Society of Brno, proposing that these traits were inherited independently as discrete particles of information (later called genes) which could randomly reshuffle and separate between generations.

It was not until after Mendel's death in 1884, however, that other scientists recognized the importance of his findings; these included Dutch scientist Hugo de Vries, whose report in 1890 was later noticed by William Bateson of the University of Cambridge, UK. Bateson went on to discover, in 1904, 'linkage', in which two or more characteristics are inherited together. When breeding sweet peas, he found, for example, that purple flowers and long pollen grains often appeared together, whereas red flowers almost always had round pollen grains.

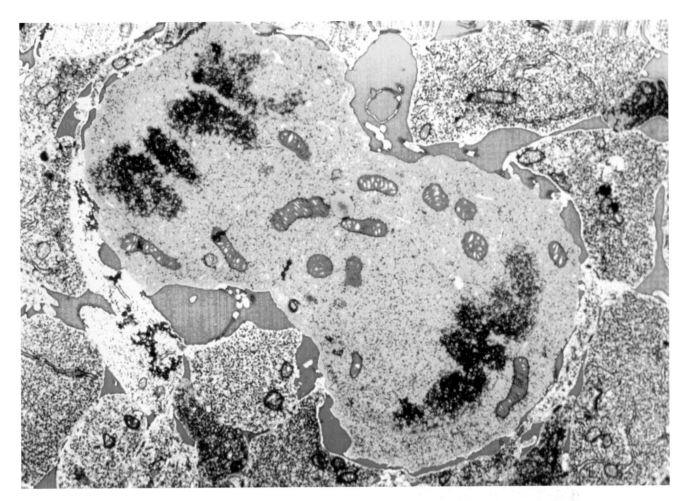

The chromosome trail

Anatomists and embryologists next took centre stage, giving Mendel's units of heredity a physical location on chromosomes, the string-like structures that can be seen under a microscope in the nuclei of dividing cells. Now we know that all cells in the body (except red blood cells, sperm and eggs) have identical sets of chromosomes, which differ in number for different species. Humans, for example, have 23 pairs of chromosomes, whereas a mosquito has just 6.

In 1890 the German scientist Theodor Boveri reported that the eggs and sperm of sea urchins each contribute an equal number of chromosomes to their offspring. In 1902 he proposed that chromosomes carried Mendel's 'factors', and that their combination in a fertilized

Above: Coloured transmission electron micrograph of a section through a dividing cell during the mitosis stage of cell division: paired chromosomes (red) separate and move to opposite poles before the parent cell divides into two new daughter cells. [BSIP / Science Photo Library]

Right: Chromosomes each consist of one long molecule of DNA wrapped around proteins called histones and coiled further into the shapes recognized under the microscope.

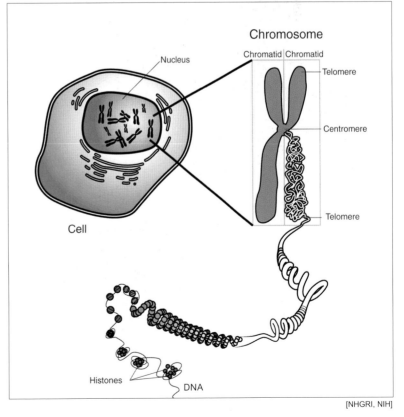

Chromosome

Nucleus

Chromatid | Chromatid

Telomere

Centromere

Telomere

Cell

Histones

DNA

[NHGRI, NIH]

Recipe for DNA

The basic building blocks of nucleic acids are called nucleotides, each consisting of a sugar, a phosphate and a base. All genetic information for a new person, animal, plant or micro-organism is stored in the sequence of the bases.

SUGAR

The sugar is a pentagonal ring of five carbon atoms linked to oxygen and hydrogen atoms. In 1909 Phoebus Levene discovered one of the sugars of nucleic acid, called ribose – giving the name RiboNucleic Acid, or RNA. In 1929 he distinguished from this the second sugar type, with one fewer oxygen atom, or deoxyribose – giving the label of DNA (DeoxyriboNucleic Acid).

PHOSPHATE

The phosphate group contains a phosphorus atom surrounded by four oxygen atoms, and confers acidity.

BASES

In DNA the four bases, consisting mainly of nitrogen and carbon atoms, are guanine, cytosine, adenine and thymine; while in RNA uracil replaces thymine.

IN 2-D

The two-dimensional arrangement of DNA involves a 'backbone' of alternating sugars and phosphates, in which the third carbon of one sugar ring links via phosphate to the fifth carbon of the next sugar and so on, an arrangement known to biochemists as 3' to 5' (pronounced 'three-prime to five-prime'). The bases are attached to the sugars – but to begin with their orientation in three-dimensional space was a mystery.

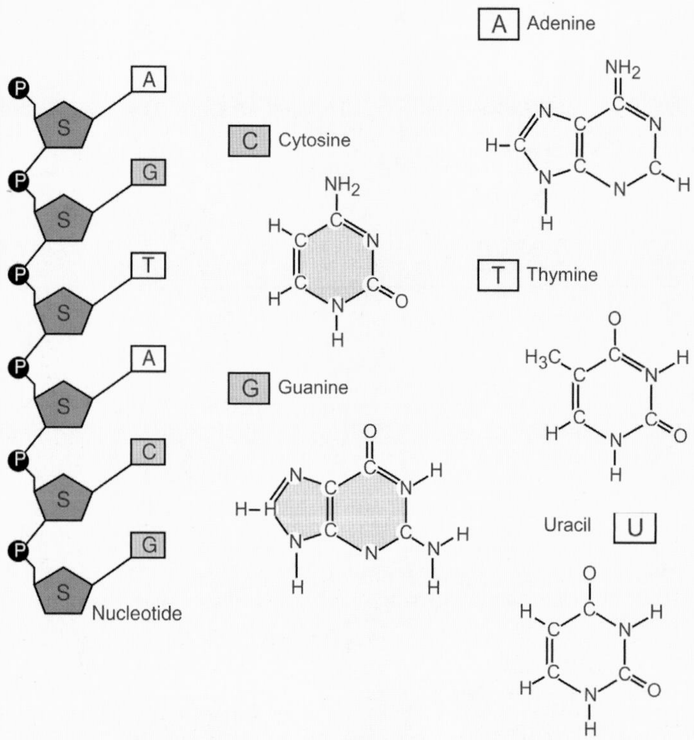

Nucleotide

A Adenine
C Cytosine
T Thymine
G Guanine
Uracil U

[NHGRI, NIH]

egg may account for the random mixing of traits. The US scientist Walter Sutton confirmed the idea independently, with the observation that chromosomes normally exist in pairs and separate from their partners during the formation of eggs and sperm (a process called meiosis). Boveri later discovered a link between abnormal numbers of chromosomes and cancer. In 1909 the Danish biologist Wilhelm Johannsen coined the term 'gene' to describe the units of heredity.

The real proof of a link between genes and chromosomes came in 1910, when American embryologist Thomas Hunt Morgan pinpointed the chromosomal locations of various genes in fruit flies, including those for wing size and eye colour, and those that determine sex on the X (female) and Y (male) chromosomes. He showed that the 'linkage' first noted by Bateson was due to two or more genes being close together on a chromosome, a key step decades later in finding genes involved in diseases such as cystic fibrosis and Huntington's disease.

Much later, in the 1950s and 1960s scientists developed a much clearer understanding of what genes are and how they function. Genes consist of one or more portions of DNA sequence spread along a chromosome which code for the production of proteins – the basic building blocks of all living cells (see Chapter 3).

Proteins versus DNA

Because chromosomes are a package of both DNA and protein, the actual substance of genes remained uncertain. Even more of a mystery was how genes dictate the appearance of a new individual. An important investigative tool appeared in 1926, with the discovery by Hermann Müller that X-rays could be used to induce mutations in fruit flies.

Back in 1869, however, unaware of Mendel and his peas, biochemists had embarked on a line that would provide an entirely different set of clues. At the University of Basel 24-year-old Johann Friedrich Miescher had used pus-soaked bandages from the local hospital – a good source of white blood cells with large nuclei – to extract an acidic, phosphorus-rich substance which he called nuclein. Indestructable by protein-chopping enzymes, nuclein was clearly distinct from the proteins of the nucleus. Later, it became known as nucleic acid. Its purpose remained a mystery.

As Horace Freeland Judson wrote in *The Eighth Day of Creation*, Miescher proposed in a letter to his uncle that large biological molecules could be responsible for carrying hereditary information: 'But Miescher's notion was

when Francis Crick arrived at the Cavendish Laboratory in Cambridge, many scientists still believed that the DNA molecule was too simple to provide all the instructions needed for an organism.

French scientist André Boivin helped DNA's case in 1948 by discovering that eggs and sperm each carry half as much DNA as other cells of the body – which agreed with Sutton and Boveri's findings that eggs and sperm only contain half the usual number of chromosomes.

In 1950, spurred by Avery, MacLeod and McCarty's findings, Erwin Chargaff at Columbia College of Physicians and Surgeons in New York set out to see if DNA varied between different organisms – to account for their different characteristics. He compared, among others, yeast and the bacteria that cause tuberculosis, and

Left: Oswald Avery. [courtesy of the Rockefeller University Archives]

fatally imprecise. The molecules he offered as examples were albumin and haemoglobin, both proteins.'

By the 1900s, nucleic acids were known to be present in all cells and to contain three ingredients, sugar, phosphate and bases, linked together to form what is now called a nucleotide (see 'Recipe for DNA'). But well into the twentieth century, many shared Miescher's assumption that 'genes' were made of protein. This was reasonable at the time, given that proteins are composed of complex combinations of up to 20 amino acid building blocks, compared to the 5 different bases known for nucleic acids: adenine, thymine, cytosine, guanine and uracil.

The ascent of DNA

Microbes provided the first opportunity to settle the dispute over whether DNA or protein was the hereditary material. In 1928 English physician Frederick Griffith found that a strain of pneumococcus, the bacteria that cause pneumonia, could pass on its disease-causing ability to a harmless strain, through a 'transforming factor'.

Sixteen years later, at the Rockefeller University in New York, Oswald Avery, Colin MacLeod and Maclyn McCarty extracted DNA from a heat-killed virulent strain of pneumococcus, injected it into a harmless strain, and found that this alone was sufficient to alter the bacteria, as Griffith had done. In preparation for the Rockefeller University's 50th anniversary celebration of the occasion in 1994, Torsten Wiesel declared, 'It is no exaggeration to say that the finding of Avery, MacLeod and McCarty opened the gateway to the modern era of biology and medicine.'

But it was still some time before their finding got the recognition it deserved. Even in 1949,

"DNA is the common thread that links every living thing with a single primaeval ancestor."

John Sulston, *The Common Thread*

found that the relative amounts of the DNA bases – adenine (A), thymine (T), guanine (G) and cytosine (C) – did vary between species. But, most importantly, he found that within a species the number of As and Ts was always equal, as was the number of Gs and Cs. These ratios later became known as 'Chargaff's rules', although no one knew how significant they were until Watson and Crick had their famous flash of inspiration.

Colin MacLeod (left) and Maclyn McCarty (right). [courtesy of the Rockefeller University Archives]

The triumph of 1953

"The chase for the double-helical structure of DNA was an adventure story in the best sense. First, there was a pot of scientific gold to be found – possibly very soon. Second, among the explorers who raced to find it, there was much bravado, unexpected lapses of reason, and painful acceptances of the fates not going well."

James Watson, *Genes, Girls and Gamow*, 2001

The year 1953 was an astonishing one in which the new partnership of physics and biology yielded one of the twentieth century's greatest triumphs: the discovery of the DNA double helix. The twisting ladder-like structure, with paired bases forming the 'rungs', was to spark a revolution in biology and medicine.

The players

Hugh Huxley recalled, 'we were young, the dark clouds of the 1930s and the war years had rolled away, and we were surrounded by exciting science, confident that the X-ray technique would yield results in the end'.

Right: Max Perutz. [PA]

Opposite page: James Dewey Watson and Salvador Luria at the 1953 Cold Spring Harbor Laboratory Symposium on viruses, June 1953. [James D. Watson Collection, Cold Spring Harbor Laboratory (CSHL) Archives]

Max Perutz

Austrian-born chemist Max Perutz, arrived at the Cavendish Laboratory in Cambridge in 1936, 'a quiet, modest and gentle man, but with the highest standards', according to his research student Hugh Huxley. Perutz had come to learn, from the masters themselves, how to use the power of X-rays to glimpse inside the crystals of biological molecules. Back in 1912, Cavendish Professor Lawrence Bragg had discovered the use of X-rays in revealing the three-dimensional structure of simple crystals, such as common salt (sodium chloride). At the age of 25 he became the youngest ever recipient of a Nobel Prize in 1915, when he won the physics prize jointly with his father, Sir William Bragg.

In 1936 Perutz published his X-ray studies of the red blood protein haemoglobin, and in 1947, according to Huxley, he created at the Cavendish 'an extraordinary successful biology laboratory'. Funded by the UK government's Medical Research Council (MRC), which wanted to see the fruits of wartime advances in physics research applied to medicine, this 'Unit for the Study of the Molecular Structure of Biological Systems' soon became a Mecca for those wishing to learn structural biology.

Vernon M. Ingram recounted in *Nature*:

Perutz attracted a nucleus of remarkably able young collaborators ... The excitement about our work was palpable; it permeated every conversation and dominated our leisure time ... we were spurred on and held together by an obsessive desire to understand the molecules of life – proteins and nucleic acids.

James Dewey Watson

James Watson was the younger of the dynamic duo who discovered the structure of DNA. He was 23 years old when he arrived in Cambridge in October 1951, scruffily dressed, noted biophysicist and historian Walter Gratzer, with a 'brash and precocious self-assurance'. Gratzer added, 'It is little wonder that Sir Lawrence Bragg, the Cambridge mandarin who found even the ebullient Crick hard to tolerate at times, did not know what to make of this strange visitor from Mars.'

In many circles, wrote Crick, Watson was regarded as 'too bright to be really sound'.

Watson once described himself as 'the only individual in Cambridge who lived solely to understand how DNA functioned as the gene'. He had developed a fascination for DNA during his PhD studies at the University of Indiana, where his mentor Salvador Luria had pursued the question of how certain viruses, called phages, could hijack bacteria until they were literally full to bursting with new phage particles. The phages injected a mixture of DNA and proteins into the bacteria, and Luria believed that it was DNA that carried the necessary blueprint of instructions for making replicas.

Watson hoped that an apprenticeship with John Kendrew, looking at the muscle protein myoglobin, would serve as a route to applying the same tools to DNA. His awareness of the importance of understanding genes was sparked by reading, as a youth, Erwin Schrödinger's *What is Life*, a text that influenced a whole generation of scientists, including Francis Crick. Watson was convinced that deciphering the structure of DNA would yield the secret of how genes were copied and passed between generations.

Francis Crick

A physicist who had helped to design land mines during the Second World War, Francis Crick had famously applied 'the gossip test' to decide what to do next in life. He realized that what he liked to talk about most was 'the borderline between the living and the non-living'. At the age of 31 he arrived in Cambridge in 1947 to do a PhD in biophysics at the Strangeways Laboratory, working on the physical properties of cytoplasm. But he soon defected to Perutz's unit to learn X-ray crystallography of proteins.

"The DNA structure initiated an intellectual revolution that has given us answers to questions that have exercised the human mind since the dawn of reason."

Walter Gratzer

Francis Crick and James Watson in Cambridge, UK.
[From *The Double Helix* by James D. Watson, Athenaeum Publishers, 1968]

physicist. During the war he had worked in atomic weapons research at the University of California at Berkeley, on isotope separation. His colleague there, John T. Randall, became the biophysics professor at King's and recruited Wilkins as his assistant. From Crick, whom he met at the Admiralty in London during the Second World War, Wilkins had learned much about DNA. His first investigation of the molecule was using ultrasound in cultured cells.

In January 1951, Wilkins was looking forward to the arrival of Rosalind Franklin from Paris, whom he thought would 'become a member of my team' (see interview with Wilkins). But this expectation was dashed as Franklin held an entirely different impression, that the X-ray diffraction of DNA was to be her territory. The ensuing tensions meant that communication between them became almost non-existent, a situation that seems to have directly influenced the subsequent outcome of the quest to discover the structure of DNA.

Rosalind Franklin

Brenda Maddox, in her biography *Rosalind Franklin: The Dark Lady of DNA*, reported that Rosalind Franklin arrived at King's College in London in January 1951 'in a state of glum apprehension'.

'I have never seen Francis Crick in a modest mood', wrote James Watson at the beginning of *The Double Helix*. From the start of their friendship, Crick was always 'the older brother' to

> "James Watson starts his book with 'I've never seen Francis in a modest mood' and the same could be said of Jim !"
>
> John Sulston, The Wellcome Trust Sanger Institute

Watson, inspiring, interpreting and batting ideas back and forth with his younger partner. The two would talk non-stop in their small office. Crick recounted in *What Mad Pursuit*, 'A certain youthful arrogance, a ruthlessness, and an impatience with sloppy thinking came naturally to both of us.'

He was already beginning to ponder the nature of the gene, and how it functioned. He knew that most of a cell's genes were on chromosomes, and that chromosomes were made of nucleoprotein (protein and DNA) and perhaps some RNA. But those around him still questioned the precise role of DNA.

Maurice Wilkins

Originally from New Zealand, Maurice Wilkins was the first scientist at the King's College Biophysics Unit in London to investigate the structure of DNA, using samples extracted from the thymuses of calves by Swiss chemist Professor Rudolf Signer.

Like his good friend Crick, Wilkins was a

Right: Maurice Wilkins at King's College, London.
[reproduced from *DNA – Genesis of a Discovery*, ed. S. Chomet, Newman-Hemisphere, London]

Rosalind Franklin in the Cabane des Evettes on a mountain holiday, France, 1950/1. [Vittorio Luzzati / National Portrait Gallery Picture Library]

A physical chemist, Franklin sorely missed the intellectually vibrant atmosphere of the Laboratoire Centrale des Services Chimiques de l'Etat in Paris, where she had spent four years developing world-class expertise in the use of X-ray diffraction to investigate the structure of coals. Unlike crystals, coals were notoriously irregular in packing, a point that was to be a significant limitation to Franklin's ability to interpret X-ray diffraction photographs – and a decided advantage to Crick.

Since childhood in London, she had developed a meticulous approach to her work, insisting on high personal and scientific standards. She was intellectually confident, enjoyed a good argument, and was determined to work in the best places. Maddox wrote that to her former colleague in Paris, Vittorio Luzzati, 'her facility for experiment was remarkable, and he quickly saw that she had "golden hands".'

Far more fragile were her relationships with those around her. A friend from her school days, Anne Piper, wrote in *Trends in Biochemical Sciences* (April 1998: 151–4), 'Although when relaxed Rosalind was far from [formidable], she was one of those very able people of great sensitivity who tend to mask their shyness with a brusque, abrupt manner. She never suffered fools gladly!'

Linus Pauling

Linus Carl Pauling, a dynamic and outspoken physical chemist at the California Institute of Technology in Pasadena, had written what many believe to be the most influential chemistry book of the century, *The Nature of the Chemical Bond*, with the conviction that chemistry could explain biological problems. Much to the

> "She told me later – she hardly ever spoke about DNA – she could have kicked herself for missing the symmetry."
>
> Aaron Klug

annoyance of Perutz and Bragg, he had already claimed a world first with his discovery of an important coil-shaped component of proteins, the alpha helix. He did this by folding pieces of paper in accordance with the rules of chemistry and X-ray diffraction patterns.

Watson said in *The Double Helix*, 'The combination of his prodigious mind and his infectious grin was unbeatable.'

Pauling first expressed his interest in DNA in 1946 with a theory that the gene might consist of two complementary strands – following earlier studies on antibodies which he knew had a close fit with their targets. Two years later,

Linus Pauling. [From the Ava Helen and Linus Pauling Papers, Special Collections, Oregon State University.]

Fortunate enough to be in the right place at the right time, Watson and Crick each brought unique experience to their endeavour. Most importantly, they talked continuously, shared expertise and revealed clues gleaned from others. It was Crick alone who understood fully the shapes and symmetries of the molecules inside crystals. Wilkins and Franklin provided the experimental evidence in the form of X-ray diffraction photographs. Franklin deduced which chemical components were on the outside of the molecule. Watson gathered clues, toyed with model pieces, and listened to the advice of Jerry Donohue – a former student of Pauling – which enabled him finally to fit correctly his cardboard cutouts of DNA bases into the space predicted by others.

Here's how it happened.

1951

'Helices were in the air'

Wilkins, assisted by student Ray Gosling, was already using X-rays to analyse stringy bundles of fibres, which they could pull out of gelatinous samples of dissolved DNA – rather like spinning a line of thread from a tangled mass of wool. Each jelly-like fibre consists of millions of tiny molecules of DNA, which pack together tightly as the fibres are pulled taut. Nowadays even school children do this exercise, using a glass rod to stir a concentrated solution of DNA and 'draw out' DNA fibres (see Chapter 9).

> "We were not being competitive, we were simply impatient to get at the truth."
>
> Francis Crick, *What Mad Pursuit*

To keep their specimens moist in the X-ray camera, Wilkins and Gosling bubbled hydrogen through water inside the sealed interior of the camera – the hydrogen kept an even atmosphere, and carried with it water vapour.

They were following in the footsteps of William Astbury and Florence Bell at the University of Leeds. In 1938 they had published in *Nature* the first X-ray analysis of DNA fibres, suggesting a helical structure. The bases, they thought, were stacked one above the other like pennies, with a space of about 3.4 Ångstrom between them. Wilkins and Gosling improved upon the quality of X-ray images, which they too suggested indicated a helical structure. Astbury had also obtained the first X-ray diffraction patterns of the alpha helix of protein but, unlike Pauling, had failed to determine the structure.

he identified 'the first molecular disease', sickle-cell anaemia, which he showed was due to a genetically determined abnormality in the red blood cell protein haemoglobin.

According to the records of the Linus Pauling Institute at Oregon State University, Pauling might have been first to discover the double helix if politics had not got in the way. Because his passport had been confiscated by the US authorities, he was blocked from attending a 1952 conference in London at which Maurice Wilkins presented evidence for DNA being helical. It was the McCarthy era of paranoia, in which his anti-weapons stance was considered a threat to the nation. His son, Peter Pauling, who arrived in Cambridge in 1952, proved to be a key go-between, relaying news between the Californian and Cambridge teams and inadvertently raising the heat of competition.

The crucial steps

How James Watson and Francis Crick deduced the structure of DNA is a drama of the most compelling kind, with personal conflict, competition and serendipity. It was heightened by the tragic early death of Rosalind Franklin at the age of 37 in 1958.

Wilkins went on the conference circuit to present the idea, taking in Naples in May 1951, and Cambridge in July of the same year. He stressed that regular patterns of spots on the X-ray photographs indicated regular packing of individual DNA molecules to form an almost crystal-like arrangement, which Wilkins called crystalline.

To Watson, who attended the Naples meeting before going to Cambridge, this was sheer inspiration. He wrote in *The Double Helix*, 'Before Maurice's talk I had worried about the possibility that the gene might be fantastically irregular. Now however, I knew that genes could crystallize; hence they must have a regular structure that could be solved in a straightforward fashion.'

Being crystalline rather than a true crystal, the DNA molecules are in clusters – or mini-crystals – within the fibres, each cluster having a different orientation. The resulting X-ray images therefore had a more diffuse and scattered array of spots than if the sample had been a crystal (see 'From A to B and back again').

Neither Astbury nor Wilkins had put forward a hypothesis for how a helical structure might enable DNA to be copied and passed on through generations, so the opportunity was there for Watson to seize.

Meanwhile Crick, already a member of Perutz's group, attended a meeting in June 1951 at which Linus Pauling triumphantly presented his new protein alpha helix. Crick was duly impressed by the usefulness of model building. 'Helices were in the air', he wrote later in *What Mad Pursuit*.

Golden hands

Rosalind Franklin arrived at King's in January 1951, already diverted from working on proteins by a letter from Biophysics Unit director John T. Randall suggesting that she switch to working on DNA. This, he had implied, would be her territory rather than Wilkins':

As far as the experimental effort is concerned there will be at the moment only yourself and Gosling, together with the temporary assistance of a graduate from Syracuse, Mrs Heller. Gosling, working in conjunction with Wilkins, has already found that fibres of deoxyribose nucleic acid derived from material provided by Professor Signer of Berne give remarkably good fibre diagrams.

Immediately, with her 'golden hands', Franklin set about improving the quality of the X-ray diffraction photographs, assembling a new camera and increasing the intensity of the X-ray beam to achieve sharper images. Franklin also discovered

Seeing is believing

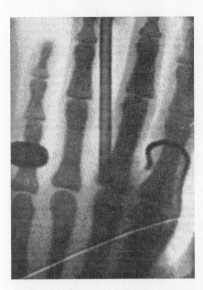

The first X-ray photograph ever taken was that of a hand – revealing how the bones and joints are positioned in space. Yet the pictures themselves were only a two-dimensional representation. Similarly, when it comes to taking pictures of fibres and crystals, they give a pretty picture of spots and lines. They need careful mathematical interpretation to deduce where the atoms lie in three-dimensional space.

X-rays, with a wavelength a thousand times smaller than that of visible light, can provide pictures of things that are too small to be seen with the naked eye, or with a light microscope. One Ångstrom is one ten-millionth of a millimetre.

Simple cameras collect the light rays that come bouncing off objects around us and focus them into an image of those objects on light-sensitive film. But there are no lenses to collect X-rays. Instead the task of forming an image has to be done mathematically from the completely unfocused pattern formed on film behind the object. This is such a difficult task that it can only be achieved for the simple case of a crystal where the same pattern of atoms is repeated over and over again, resulting in a regular pattern of spots on the film whose intensity can be accurately measured. (Nowadays, instead of film, researchers use arrays of light-sensitive meters connected to computers to measure and analyse the X-rays, a set-up less like a camera and more like our eyes in which an image forms on an array of photoreceptor cells connected to our brains.) (See Chapter 5.)

Crick knew from recent X-ray crystallography studies of proteins that shape and function of a molecule are intimately linked. In a protein crystal, all molecules had to have the same shape to give a clear X-ray diffraction pattern. And losing the shape – through boiling, for example – causes a loss of function.

Only in the early 1980s, with the advent of recombinant technology allowing the large-scale production of short stretches of DNA, could enough short stretches of DNA be isolated in pure form to grow crystals, and confirm the Watson and Crick model. Alex Rich at the Massachusetts Institute of Technology did this, and discovered a new form of DNA – Z DNA – which only occurs transiently during unwinding of the helix.

that she could persuade DNA fibres to become longer and thinner, producing yet clearer data. To do so she compared different relative humidities (water content of air) inside the camera, by bubbling hydrogen through a series of salt solutions of different concentration.

At 75 per cent humidity, the DNA sample was in the crystalline state described by Wilkins, which she called A form. By increasing the humidity to around 92 per cent, Franklin found

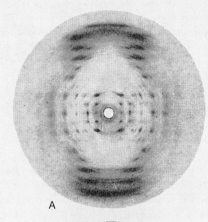

that the sample quickly became wetter and easier to pull into thinner threads. This she called B form or paracrystalline DNA, meaning that the rigid mini-crystal state was broken up into a looser arrangement.

Comparing X-ray images of both A and B forms, Franklin saw that B-form DNA gave clearer patterns – what is now referred to as the 'helical cross'. She also saw that individual DNA molecules in crystalline DNA must be held together by weak physical forces, easily broken as water floods in; the DNA molecules get pushed apart, and escape the tight packing arrangement. As they float around more freely, the molecules of B-form DNA are more amenable to being teased into longer, thinner fibres. But ironically, this also encourages the molecules to line up in a more uniform alignment, even though their orientation is still random – hence the clear images of the B form (see picture in 'From A to B and back again'). (See *Nature*, 219: 808–11, 1968.)

Ironically, because rather than focusing all her efforts on B-form DNA, Franklin appar-

Rosalind Franklin, Lyons, 1949. [courtesy of Rachel Glaeser]

From A to B and back again

perfect crystalline array:
all molecules in same orientation

disordered array:
molecules in random orientations
(B-form DNA fibre)

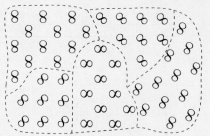

small crystalline blocks in random orientations
(A-form DNA fibre)

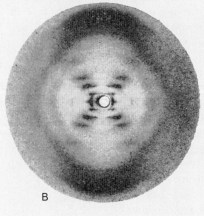

Left: Rosalind Franklin's X-ray diffraction patterns of the A and B forms of DNA. [reproduced from *Acta Crystallographica*]

Above: Orientation of DNA molecules in A and B forms of DNA fibres. See 'Golden hands' for details. [courtesy of Aaron Klug]

ently directed her attention for many months on the A form, wishing to carry out crystallographic analysis. Later commentators have suggested that this decision only served to delay her reaching a firm conclusion about the structure of DNA being a helix.

In May of the following year, 1952, Franklin took her best X-ray image of B-form DNA, little knowing that it would later become famous as a lightening bolt of inspiration for Watson.

On the outside

Nevertheless, Franklin gained the critical insight that the surface of DNA must have attraction for other DNA molecules in a way that is strong enough to hold them together in a crystalline arrangement, and yet weak enough to be smashed by water. The only chemical components of DNA that fit the criteria were the phosphates in the 'backbone'. These, therefore, were on the outside of the DNA molecule. The PhD thesis of a Norwegian student, Sven Furberg, was circulating round the department at King's, correcting Astbury's assumption that the bases of DNA lay parallel to the sugars. Instead, Furberg surmised, the bases were perpendicular to the backbone, which to Franklin meant that they pointed inwards away from water.

In the meantime, completely unaware of Franklin's letter of appointment from Randall, Wilkins assumed that he was at liberty to speak on behalf of the King's team. This misunderstanding boiled over into an ugly confrontation in July 1951 when Wilkins presented King's data on DNA at the Perutz unit. According to Maddox's biography, Franklin was unable to contain her anger at what she saw as an invasion of her turf, and publically rebuked Wilkins:

> Her work was her life, the core of her identity. Undervalued at King's, she had just achieved extraordinary results by working in virtual isolation. Now what she saw as a less able colleague of higher rank was proposing to elbow in and spoil the clarity of her investigation.

Wilkins later described how he obtained 'basically similar patterns from DNA from various sources and from sperm heads' in early 1952, enabling him to confirm his own conviction about a helical structure (*Science*, 27 June 1969: 1537–9). But again this result gave no insight into how genes are copied.

Watson comes bounding in

In October 1951, Watson joined Perutz's lab, sharing an office with Crick, where they immediately discovered their common ground – an interest in solving the structure of DNA. Both were ambitious and impatient, but neither was officially working on DNA. Crick introduced Watson to Wilkins a month or so later, on one of Wilkins's increasingly frequent visits to Cambridge (apparently seeking solace and encouragement). Watson and Crick, however, remained coy about their own interest in solving the structure of DNA; Watson was ostensibly doing X-ray studies of plant viruses. He summed up his impression, that Wilkins and Franklin 'were set to provide the definitive evidence for choosing one DNA model over another. But over the next year, their personalities clashed badly, and Maurice found himself driven away from X-ray analysis of DNA.'

In November 1951 Watson attended a colloquium at King's at which Franklin presented her data on the A and B forms of DNA. But, as was his habit, Watson failed to take notes and missed her vital clue that the phosphates lay on the outside of the molecule.

Thinking that they were on the right tracks, Watson and Crick made a blundering attempt at building a model of DNA, proposing three chains of nucleotides per molecule, each with the phosphate backbone lying in the centre.

> "DNA is, at bottom, much less sophisticated … it was just good luck that we stumbled onto such a beautiful structure."
>
> Francis Crick, *What Mad Pursuit*

Franklin wasted no time in pointing out their error regarding the phosphate backbone. Watson wrote later in *The Double Helix*: 'the embarrassing fact came out that my recollection of the water content of Rosy's DNA samples could not be right'.

Furious that his associates had been ridiculed, Bragg told Watson and Crick to leave DNA to the King's group to solve. Both Watson and Crick have since emphasized that their next move was to offer to hand their jigs over to Wilkins, encouraging him to pursue model building instead.

1952

The evidence

In February 1952 Franklin prepared a report on the previous year's work, including a description of the unit cell – the repeating portion – of crystalline (A-form) DNA. Using the International Tables of Crystallography, she had assigned its symmetry as 'face-centred monoclinic'.

Unfortunately, owing to lack of experience

Facsimile of Rosalind Franklin's notes showing her attempts to deduce the symmetry of the phosphate-sugar backbones of the A-form of DNA. [courtesy of Jenifer Glynn / source: Churchill College Archives, Cambridge]

in dealing with regular crystals, Franklin overlooked this as a two-fold symmetry of the unit cell – in other words, that after rotating the DNA molecule through 180 degrees about a central axis it would look the same. The significance of symmetry dawned only on Crick much later in the story.

She also concluded that DNA is likely to be helical '(which must be very closely packed) containing probably 2, 3 or 4 co-axial nucleic acid chains per helical unit, with the phosphate groups near the outside'.

This description appeared again as part of a formal report to the MRC in December 1952, distributed to all members of the relevant supervising committee, including Perutz.

The Paulings enter the fray

In December 1952 Linus Pauling wrote to his son Peter, now also studying in Perutz's unit, to say that he and his colleague Robert Corey had devised their own model of the structure of DNA. The letter spurred Bragg and Perutz, still smarting from their previous humiliation by Pauling, into allowing Watson and Crick to resume DNA model building. Anxious not to lose time, Watson began model building using cardboard pieces – while waiting for the metal workshop at Cambridge to supply parts.

1953

In January 1953 Pauling sent the paper describing his DNA model to the Cavendish Laboratory, where Watson claimed immediately to have spotted the same fatal error that he made in the summer. Pauling was proposing a triple helix of three chains of nucleotides

twisted round each other, with the phosphate backbone lying in the centre, and no sign of the known acidic properties of the bases. Jubilant, Watson visited King's, and told Franklin about the paper. She, according to Watson's account in *Genes, Girls and Gamow*, dismissed the notion of a helical structure (but, as Aaron Klug later emphasized, her doubts related only to the A form of DNA; she had already noted in her lab books that the B form was helical):

Rosalind, however, thought I was being unnecessarily hysterical, telling me in no uncertain terms that DNA was not helical. Afterwards, in the safety of his office, Maurice – bristling with anger at having been shackled now for almost two years by Rosalind's intransigence – let loose the, until then, closely guarded King's secret that DNA existed in a paracrystalline (B) form as well as a crystalline (A) form. In his mind the cross-shaped B-diffraction pattern, shown on the X-ray he then impulsively took out of a drawer for me to see, had to arise from helical symmetry.

It was a jaw-dropping experience for Watson: a stunning sign of a helix. Watson returned to Cambridge by train, and in a rare moment of record-keeping he sketched the DNA helix from memory in the margin of his newspaper. According to Crick's recollections in *Nature* in 1974, Watson 'certainly had not remembered enough details to construct the arguments about Bessel functions and distances which the experimentalist gave. I myself, at that time, had not seen the picture at all. Consequently we were mildly surprised to discover that they had got so far and delighted to see how well their evidence supported our idea.'

Perutz's hand of fate

In February 1953 Perutz showed Watson and Crick the MRC report, an action which he later defended on the grounds that the report was not confidential and that they had asked to see it. Perutz acknowledged later that it did help Crick to realize the significance of Franklin's assignment of the symmetry of unit cell as 'face-centred monoclinic' (but that he could have known this earlier if Watson had been better at note-taking). To Crick this was the same symmetry assignment that he was analysing for haemoglobin. Going the extra step, Crick suggested therefore that DNA contains two strands of nucleotides, coiled into a helix, running in opposite directions, one up, and one down. (The directions, according to the orientation of the chemical bonds between sugars and phosphates, are 3' to 5' for one strand, and 5' to 3' for the other (see Chapter 1).)

Watson admitted in *The Double Helix* that 'no one at King's realized [the data] were in our hands ... as soon as Max saw the sections by Rosy and Maurice, he brought the report in to Francis and me'.

There remains some dispute over how important the MRC report was in revealing to Watson and Crick that the DNA helix had a diameter of 20 Ångstroms. This information helped Watson to know the space into which the bases of DNA had to be squeezed. Watson maintains that he had this information already from the work of Wilkins and Astbury (see interview on pages 39–42, Chapter 4).

Base pairing

Watson tried to get the cardboard cutout bases to fit into a 20 Ångstrom diameter helix without causing it to buckle in or out. With two chains in parallel – as in the sides of a ladder – the bases, if at right angles, could meet in the middle to form the 'rungs'. But two purines (adenine and guanine) paired together would mean a gap in the middle – the rungs would be broken – while pyrimidines together (thymine and cytosine) would distort the 'rungs'.

While Watson fiddled on, he was also trying to solve the riddle of Chargaff's rules. Chargaff had visited them in Cambridge and reiterated his findings that adenine and thymine are in a 1:1 ratio, as were guanine and cytosine. But before Watson could find the answer, a lucky intervention came from Jerry Donohue who shared the office with Watson and Crick. He suggested that Watson use an alternative chemical arrangement of the bases, in which a hydrogen atom shifts to a different carbon. It meant

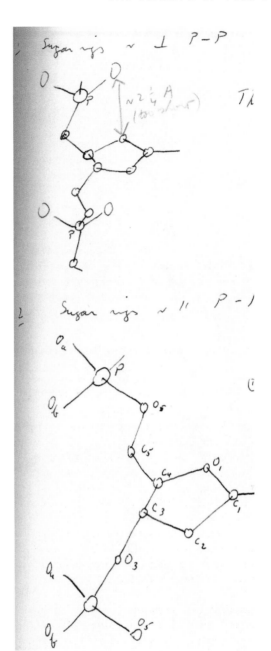

Rosalind Franklin's notes about the relative positions of phosphates and sugars in DNA for crystallographic analysis. [courtesy of Jenifer Glynn / source: Churchill College Archives, Cambridge]

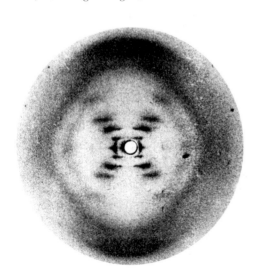

Left: Rosalind Franklin's 'Photograph 51' of B-form DNA as it appears in *Nature*, 25 April 1953.

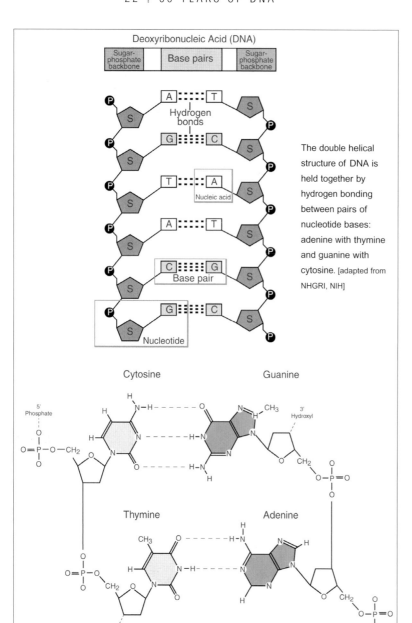

The double helical structure of DNA is held together by hydrogen bonding between pairs of nucleotide bases: adenine with thymine and guanine with cytosine. [adapted from NHGRI, NIH]

using the 'keto' instead of the 'enol' form because the 'enol' bases would not have formed neat pairs.

In *Nature*'s 21st anniversary celebration, in 1974, of its publication of the papers describing DNA's double helical structure, historian Robert Olby wrote that Watson and Crick speeded up the general acceptance of Chargaff's findings, which they learned:

from Chargaff himself when they met him in Cambridge in 1952. Some nine months after that meeting they published their famous model in Nature. *The key to the model was the pairing of the bases, and this was done in accordance with the Chargaff rules – adenine with thymine and guanine with cytosine, so that A/T = G/C = 1. Now, 21 years later, what a rich harvest has been reaped from that neat idea!*

Franklin juggles
On 23 February 1953 Franklin recorded in her notes that both A and B forms were likely to be two-chain helices. And in an attempt to explain Chargaff's rules, she proposed that the bases adenine and thymine may be interchangeable, and guanine and cytosine likewise. But she did not get as far as envisaging base pairing.

Eureka!
Watson repositioned the cardboard pieces so that adenine paired with thymine and guanine with cytosine. Eureka! Both pairs had the same overall dimensions, and fitted neatly into a double stranded helix of 20 Ångstrom diameter. The bases were 'stuck' in their pairs with a 'glue' of hydrogen bonds, holding the two phosphate-sugar chains parallel along their entire length. Crick saw that the chains ran in opposite directions, and that the model satisfied Chargaff's rules. The real brilliance of the model was in revealing how base pairing

Aaron Klug annotated Rosalind Franklin's notebook in 1968 to say 'REF is at last making the correct connection between the A and B' [courtesy of Jenifer Glynn / source: Churchill College Archives, Cambridge] enables each of the two chains to 'unzip' and serve as a template, or mould, for the formation of another complementary chain. AGGCAT, for example, would direct the joining of TCCGTA.

The model showed the structure of DNA to be 'far more beautiful that we ever anticipated', Watson wrote in *Girls, Genes and Gamow*. It was 'so perfect that the experimental evidence in its favor from King's almost seemed an unnecessary accompaniment to a graceful composition put together in heaven'.

Crick went straight to their favourite public house, the Eagle, and announced to anyone bothered to listen that they had 'found the secret of life' (*The Double Helix*).

Flip of a coin

On 2 April Watson and Crick sent to *Nature* their paper describing their model. They flipped a coin to decide who would be first author and opened with the cautious statement: 'This structure has novel features which are of considerable biological interest.' They concluded with the most famous scientific understatement of all time: 'It has not escaped our notice that the specific pairing we have postulated immediately suggests a possible copying mechanism for the genetic material.'

In *Nature*'s 21st anniversary celebration issue, 26 April 1974, Crick admitted:

> this has been described as 'coy', a word that few would normally associate with either of the authors, at least in their scientific work. In fact it was a compromise, reflecting a difference of opinion. I was keen that the paper should discuss the genetic implications. Watson was against it. He suffered from periodic fears that the structure might be wrong and that he had made an ass of himself.

Franklin had already drafted her manuscript on 17 April, before knowing about Watson and

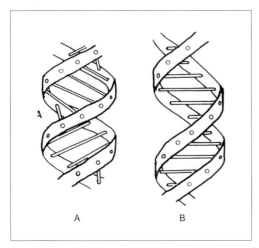

A   B

Crick's paper. It described X-ray photos of the B form of DNA revealing as 'highly probable' a helical structure, with the phosphate groups lying on the outside of a double helix, a diameter of 20 Ångstroms, the two chains separated by a distance of 13 Ångstroms.

A hat trick

On Friday 25 April 1953 three papers appeared in *Nature*, one from Watson and Crick, one from Wilkins and his colleagues, and one from Franklin and Gosling. Wilkins' paper described a 'system of helices (corresponding to a spiral staircase with the core removed)', and reported similar X-ray diffraction patterns in DNA from different species and cell types: trout semen and

Replica of Watson and Crick's brass plates model of DNA, on display at the Dolan DNA Learning Center, CSHL. [courtesy of Dolan DNA Learning Center, CSHL]

Left: Diagram of the A and B forms of DNA; the molecules of the A structure (left) are 30 per cent shorter, tightly packed in microcrystals, with 11 nucleotides per helical turn of 28 Å. In the B structure (right), there are 10 nucleotides per helical repeat of 34 Å. [*Nature*, 219: 809, 1968]

Rosalind Franklin flanked by colleagues (left to right): Francis Crick, Don Caspar, Aaron Klug, Rosalind Franklin, Odile Crick and John Kendrew in Madrid, 1956, at an International Union of Crystallography meeting.

[courtesy of Don Caspar]

phages (bacterial viruses). Later in July Franklin and Gosling published in *Nature* a paper showing that the A form also contains two-chain helical molecules similar to the B structure. In A-form DNA each chain contains 11 nucleotides per helical repeat of 28 Ångstrom, while in B-form DNA there are 10.

The aftermath

The events of 1953 produced a torrent of questions about the manner in which Watson and Crick made their discovery.

The Nobel Prize

Watson, Crick and Wilkins shared the Nobel Prize for Physiology or Medicine in 1962, four years after Franklin's death. Only Wilkins acknowledged that Franklin made 'very valuable contributions to the X-ray analysis'. The rest of the world knew little of the events of 1953 until Watson published his phenomenal bestseller *The Double Helix* in 1968.

This account, written at a popular level, gave the first glimpse into the strained relations between Franklin and Wilkins, the manner in which Watson came across Franklin's Photo 51, and how the MRC report came into Watson and Crick's hands. It caused an uproar – not least because of Watson's constant references to Franklin as 'Rosy', which no one used to her face, as well as his derogatory remarks such as: 'Clearly Rosy had to go or be put in her place. The former was obviously preferable because, given her belligerent moods, it would be very difficult for Maurice to maintain a dominant position that would allow him to think unhindered about DNA.'

Watson stated that his intention was to recreate his first impressions rather than to be objective, to 'convey the spirit of an adventure characterized both by youthful arrogance and by the belief that the truth, once found, would be simple as well as pretty'.

But many have taken exception to his portrayal of Franklin – including her friend Anne Sayre, who wrote her own account of Franklin's role, creating her as a legendary feminist icon. More recently, Brenda Maddox redresses the balance to emphasize Franklin's skills and expertise as a scientist (see essay on page 93). Maddox also reinforces the view that Franklin's data were 'fundamental to the discovery'.

'Those who have even the slightest acquaintance with Franklin know that the criticisms of her that permeate *The Double Helix* do not reflect the real person', wrote David Blow, a researcher in the Biophysics Group at Imperial College of Science, Technology and Medicine (*Nature*, 418: 725–6, 2002).

How close was Franklin?

Aaron Klug, Franklin's close friend and colleague, recollects: 'On the basis of those notebooks I realized how close she came to discovering the structure. There's no doubt in my mind that left to her own devices she soon would have got it out.'

The problem, according to Watson, was that Franklin was too much in isolation: 'If she had talked to Francis, and he'd seen the pictures, he would have told her how to interpret the data and she would have found the answer all by herself. She just needed help, because she had never done these things with crystals, so she had never been brought up on space groups.' (See interview on page 45, Chapter 4.)

Others predict that if Watson and Crick had not discovered the structure, then it would have 'dribbled out in bits and pieces' with far less impact.

Walter Gratzer wrote in his introduction to *Passion for DNA*:

> There would have been no comparable theatrical coup, no sudden illumination of the biological landscape that needed both the allure of the double helix (so striking a symbol that it was purloined for one of his paintings by Salvador Dali) and the implications of reciprocity between the strands. The whole was greater than the sum of its parts, and it was the manner as much as the substance of the discovery that made the impact and at once implanted a Nobel Prize in the womb of time.

Crick wrote in *Nature*'s 21st anniversary special in 1974: 'If Watson had been killed by a tennis ball I am reasonably sure I would not have solved the structure alone, but who would?' He went on to speculate that Pauling might have got there if he had realized quickly that his own model was wrong, and said Franklin was 'only two steps away', needing to see that the two chains run in opposite directions and the bases are paired. He added, 'I doubt myself whether the discovery of the structure could have been delayed for more than two or three years.'

Linus Pauling, on the other hand, was 'astonished' that he overlooked a previous discussion with his colleague Max Delbruck about genes consisting of two complementary molecules, further commenting: 'Nevertheless, I myself think that the chance is rather small that I would have thought of the double helix in 1952, before Watson and Crick made their great discovery.'

Mad hatters at the DNA tea party

Jan Witkowski, director of the Banbury Center, Cold Spring Harbor Laboratory, described in 2002 the discovery of 'the only contemporary account by an observer of events surrounding the discovery of the double-helical structure of DNA' in the archives of Rockefeller University. Witkowski identified the author of the memorandum as Gerald Roland Pomerat, assistant director of the natural sciences programme at the Rockefeller Foundation, who was inspecting the Cavendish Laboratory in April 1953. The most exciting research, he reported, was from 'two of the younger men', Watson and Crick. Witkowski continued, 'Watson tells us in *The Double Helix* that Francis Crick's excitement grew each day as he regaled visitors with the wonders of the structure. Now, Pomerat's words give us a new view of what the visitors saw of Watson and Crick: 'Somewhat mad hatters who bubble over about their new structure' (*Nature*, 415: 473–4, 2002).

Mad Hatter's tea party from *Alice's Adventures in Wonderland*, Lewis Carroll, Macmillan.

The golden years of molecular biology

"I saw the double helix as the culmination of almost a century of genetics, but for Francis it was to be the splendid beginning of a new life not only for him but for biology itself."

James Watson, *Passion for DNA: Genes, Genomes and Society*

Watson and Crick's model of the structure of DNA opened the door to discovering how genes are copied and passed from parent to offspring, and how they direct development from embryo to adult.

When they discovered the complementary base-pairing of the DNA double helix, Watson and Crick realized that genetic information contained in the sequence of bases could be copied, with one strand of DNA forming the template for the making of a new strand (DNA or RNA). The information could then be transferred from the nucleus to the cytoplasm to instruct the making of proteins. These 'workhorses' make up the architecture of cells and tissues, and carry out vital tasks such as energy uptake and use, hormone synthesis, and sending and receiving of messages.

Over the next five decades, molecular biolo-

gists were to elucidate the mechanisms involved and to make enormous advances in biology, genetics and medicine. They now have a vast tool kit for manipulating and cloning genes, producing pure proteins on an industrial scale, and 'reading' DNA sequences, ultimately to understand the full set of information stored in the 'books of life' – the genomes of entire organisms.

Right: *Bronze Helix* by Charles Reina in the Grace Auditorium, CSHL. [Julie Clayton]

"the most important thing a gene could do would be to direct the synthesis of a protein, probably by means of an RNA intermediate"

Francis Crick, *What Mad Pursuit*

TIMELINE OF MAJOR EVENTS

Life begins in the 1950s

1955
Fred Sanger deciphers the first amino acid sequence of a protein – the hormone insulin.

1956
Paul Zamecnik and Mahlon Hoagland discover the existence of 'transfer RNA', a small type of RNA molecule that transports individual amino acids to ribosomes, the protein-making machinery of cells, for assembly into proteins. Originally calling their find 'soluble RNA', Zamecnik and Hoagland identify it by trying to assemble all the parts necessary for making proteins in a test tube.

1957
Francis Crick and George Gamow propose the 'Central Dogma': a one-way flow of information from DNA in the nucleus to proteins in the cytoplasm, possibly via an RNA intermediary. They also put forward the 'sequence hypothesis', that the sequence of bases in DNA itself specifies the sequence of amino acids in protein. RNA is a single-stranded molecule of nucleic acid, similar in composition to DNA except that one of the bases, thymine, is replaced by another called uracil.

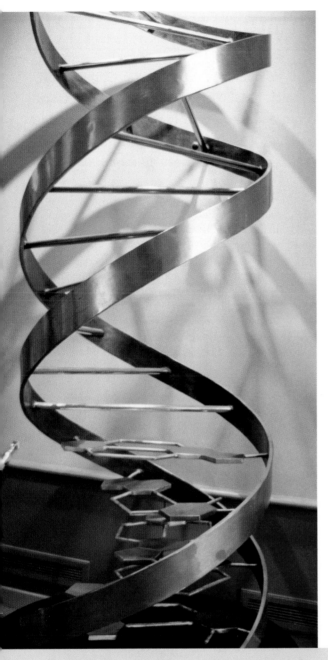

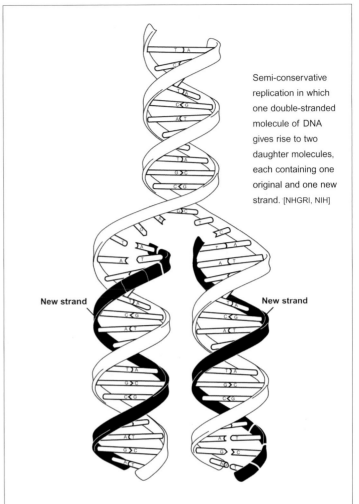

Semi-conservative replication in which one double-stranded molecule of DNA gives rise to two daughter molecules, each containing one original and one new strand. [NHGRI, NIH]

New strand

New strand

"The experiment of semi-conservative replication was absolutely key – that's what really swung a load of biologists … Otherwise X-ray diffraction was too esoteric a method for anybody to believe."

Aaron Klug

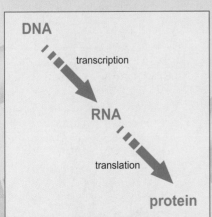

DNA

transcription

RNA

translation

protein

1957

Arthur Kornberg and Severo Ochoa discover DNA polymerase I, the first of what is now known to be an entire complex of enzymes involved in the process of DNA copying, or 'replication'. DNA polymerase I makes a new DNA strand by joining individual bases as directed by a single-strand DNA template. This is the first demonstration of DNA synthesis in a test tube, and obeys Chargaff's rules (A with T, and G with C). In 1959 Kornberg receives a Nobel prize for his work, and eight years later succeeds in being the first to create a live virus in the test tube.

1958

Matthew Meselson and Franklin Stahl provide the experimental proof for Watson and Crick's prediction in 1953 of how DNA is copied. Using DNA from the bacterium *E. coli*, they showed that the two strands of a DNA double helix separate and serve as a template for the synthesis of two 'daughter' strands. The result is two daughter helices, each containing one 'old' and one 'new' strand of DNA. They dubbed the process 'semi-conservative' replication.

INTERVIEW WITH WALTER GILBERT

A physicist drawn to the excitement of molecular biology, Walter Gilbert (now Principal Investigator of the Bioinformatics Group at Harvard University) talks here about the dramatic transition from a world where ideas were spread by itinerant scientists between just a handful of labs, to one where hundreds of labs churn out DNA sequences at a rate that doubles every 18 months.

How did you first get involved with Jim Watson?
I met Jim in the fall of 1955 at a party [in Cambridge, UK]. We came back to Harvard at the same time in the fall of '56 … and Celia my wife worked as Jim's first lab assistant. We were feeding Jim every other night – he'd come by for dinner because we lived right near the lab.

At the end of the '50s Jim's lab was working on ribosomes and hoping to find something on protein synthesis. It led to this messenger RNA

> "The early '60s were a period of very great effervescence in molecular biology."
>
> Walter Gilbert

question: 'Is there some very short-lived RNA in the cell that is made and broken down in a few minutes? And is it different from ribosomal RNA (a stable RNA)?' In the late spring [1960] Jim said to me 'There's something very exciting going on in the laboratory.' I came over and saw Jim and Francois Gros trying to do experiments on messenger RNA. I followed them around one day and the next day I joined in doing the experiments. We worked on that all summer and fall. The paper was published back to back

with the Bretscher, Cohen and Meselson paper in *Nature* in the spring of 1961.

In the summer of '61 we went to the Moscow meeting and heard Marshall Nirenberg speak in a tiny room. But actually I met Marshall at breakfast on the porch of the hotel and learned the story about poly U and the genetic code – this was the real stuff, it was very dramatic. It's the sort of thing that turns up at a meeting because you're having breakfast with somebody and you say 'What are you doing?' and he tells you this wonderful story.

The striking difference between then and now is the simplicity of the field and how it's grown. So that at that time you worked on bacteria and viruses, genetics and biochemistry – it's all part of the same field. As we started doing the messenger RNA experiment I could read six papers and know the entire background.

What was Jim like to work with?
We both showed a fascination for new and interesting results in science. I always describe Jim as having a wonderful intuition as to what the important lines of scientific research were.

Around 1964 I came up for tenure and was promoted in biophysics. At that point I began officially to run a laboratory together with Jim

TIMELINE OF MAJOR EVENTS

Code breakers of the 1960s

1960
François Jacob and Jacques Monod demonstrate the first evidence that genes can control others. The 'regulatory' genes produce proteins that 'switch' genes on or off by binding to other nearby stretches of DNA. This paves the way for the discovery of a whole slew of control proteins known as transcription factors, promoters, silencers and enhancers. The control of gene action is what determines the specialized functions of cells in time and space, in any part of the body.

1961
Sydney Brenner, François Jacob and Frank Meselson and independently Watson and Gilbert discover messenger RNA, solving the conundrum about how instructions in the nucleus, contained in DNA, can reach the cytoplasm, where proteins are made. They determine that mRNA first forms in the nucleus as a single strand on a DNA template by a process called 'transcription', then gets shuttled out of the nucleus to the surrounding cytoplasm and attaches to the cell's protein-making factories called ribosomes, where it is 'translated' into a string of amino acids.

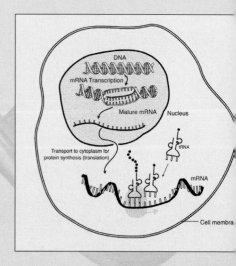

Walter Gilbert.
[Jane Reed/Harvard News Office/
©2003 President and Fellows of
Harvard College]

and had a large group of students. Jim was not a great experimenter but had a very clear conception of what the experimental path might be. I was a very competent experimenter, but I also had a vision of what the outcome might be.

The early '60s were a period of very great effervescence in molecular biology. It was just a group of people in a small number of laboratories, constantly in contact with each other and talking about the novel experimental results. In that late '60s Francis [Crick] would have been almost a Johnny Appleseed, travelling round from laboratory to laboratory, picking up information here, and dropping it at the next one. Physicist Leo Szilard also did a lot of that in the early '60s and was very influential. He visited Paris and found them talking about genetic

control mechanisms. I think he actually suggested to Jacob and Monod that there might be a negative control.

The discovery of bacterial genetics was dramatic. The impulse was to do genetics on the smallest living thing – following Monod's dictum that what's true for *E. coli* is true for elephants. Today we don't do it quite that simply … We later learned that genes in elephants have introns and exons, and genes in bacteria don't. That's a big difference, as well as splicing patterns in RNA, small interfering RNAs, combinatorial genetic controls, and much more complicated RNA polymerases. People struggled with trying to work with higher cells before molecular biologists got the power that they have today and that power wouldn't have

"The importance of the concept of the principles of genetic decoding, proved 40 years ago, remains huge."
John F. Atkins and Raymond F. Gesteland
(*Nature*, 414: 693, 2001)

1961
Sydney Brenner and Francis Crick reveal how the 'Genetic Code' works, with the sequence of DNA bases spelling out the order of amino acids in a protein. A triplet of bases, for example CAT, forms a 'word' or codons, and specifies a single amino acid. The code is 'read' in a linear sequence from a fixed starting point without 'punctuation', to the end of the 'reading frame'. The code has a built-in flexibility, with 61 out of 64 different possible triplets coding for the 20 known amino acids (from combinations of the four bases, 4x4x4x4). For example, CGC, CGA and CGG all code for arginine. They also suggest that the function of transfer RNA is to 'translate' the code from the messenger RNA intermediate.

1961
Marshall Nirenberg at the US National Institutes of Health begins 'cracking' the genetic code when he finds that a chain of mRNA consisting of just the nucleotide base uracil (poly-U) codes for the synthesis of a protein containing only phenylalanine. Five years later his team, including Har Gobind Khorana, at the University of Wisconsin, have identified the entire set of 61 codons that specify the 20 known amino acids. Three codons act as stop signals for the growing protein (TAA, TAG and TGA).

1967
Walter Gilbert and Mark Ptashne independently isolate 'regulatory proteins' predicted by the experiments of Jacob and Monod, which can switch genes on and off by binding to flanking 'control' regions of DNA.

Opposite page: messenger RNA, synthesized on a DNA template, moves from the nucleus to the cytoplasm, where it associates with ribosomes to direct protein synthesis. [NHGRI, NIH]

been achieved had one just stayed with higher cells. It was achieved by the work with phages and bacteria.

How did you develop DNA sequencing?

The transformation in the middle '70s from a world in which you could not sequence the DNA to a world in which it became extremely easy were the discoveries of Allan Maxam and myself and Alan Coulson and Fred Sanger, of finding rapid ways of displaying DNA sequence. Fred was following a particular logic – a profound belief that you should know the primary structure.

In my case I was interested in how DNA makes proteins, and how you control DNA. Allan Maxam and I worked out approximately 20 bases of the operator in two years – one base a month. It was the first longish DNA sequence to be published. In 1972 we did mutations in the operator to see whether it would block the interaction with protein. It worked so clearly that I not only could see that the repressor touches DNA just at certain bases, but I got a pattern on the gel that was so distinctive it told me where the adenines and guanines were in the sequence. I didn't have to know the sequence.

Fred and I simultaneously discovered that gels would work in this particular way and give us this fantastic reading. We made the DNA by breaking the fragments open using a chemical reaction. Fred made the fragments by synthesizing the DNA out to a certain base and then stopping it. Suddenly there was this burst of activity. We simply gave the method away at a Gordon conference and people began sequencing all over the world almost immediately. Both methods were used almost equally until the machines came in during the mid-80s, when Fred's methods lent themselves more to a kit.

During the 1970s, what did you think would be the power of recombinant DNA technology?

Those of us who got close to the technology thought it was going to be of tremendous medical and practical importance, because there were proteins in small amounts in human bodies that we could work on as therapies. That's why we chose insulin – we knew that there was a shortage … By 1979 we thought of making interferon and hepatitis vaccine. The world transformed, so that by '78/'79 there were maybe hundreds of thousands of DNA bases sequenced, but not yet millions. The amount of DNA sequenced has gone up a factor of 10 every five years since 1975. We're now at about 10 billion bases of DNA sequenced. Five years ago we were at about a billion, 5 years before that we were at 100 million. It goes up fantastically rapidly … doubling every 18 months since the discovery of the sequencing methods in '75 through to the Genome Project.

Were you always a great proponent of the Human Genome Project?

In 1985 when Bob Sinsheimer and Tom Baker pulled together a meeting to ask 'Could one sequence the human genome and should one sequence it', I went in thinking 'This is absurdly more complicated than we could ever do.' And then I realized that it was just a pure application on an appropriate scale of current technology. Can I contemplate spending 100 million dollars per year working on this problem? Of course, in the pharmaceutical industry we do this all the time … I was a strong proponent of the idea that here would be a resource that we can make, an accurate human sequence, and once we make it, we will then mine that resource, for medicine and for biology for the next hundred years.

> "Allan Maxam and I worked out approximately 20 bases … in two years – one base a month."
>
> Walter Gilbert

TIMELINE
OF MAJOR EVENTS

The cut and paste of the 1970s

1970

Howard Temin and David Baltimore independently isolate a viral enzyme, 'reverse transcriptase', which can reverse the initial flow of information proposed by Crick by making DNA from an RNA template. This enables certain viruses that store their genes as RNA, such as influenza and HIV, to insert as DNA copies into host cell DNA.

1971

Stanley Cohen, Annie Chang and Herbert Boyer use so-called restriction enzymes to perform the first 'cut and paste' of DNA from one organism to another, inserting DNA conferring antibiotic resistance into laboratory-bred bacteria. This marks the start of the biotechnology industry. In 1978 Boyer produced the first recombinant DNA drug – human insulin – in *E. coli* bacteria. Today, more than 3000 restriction enzymes have now been isolated from bacteria for use in molecular biology, where their normal role is to chop up the DNA of invading viruses.

1975

Allan Maxam, Walter Gilbert, and Fred Sanger invent manual methods for DNA sequencing. (see interview with Walter Gilbert)

A MATTER OF LIFE AND DEATH: INTERVIEW WITH KARY MULLIS

In 1985 Kary Mullis, currently Vice President of Molecular Biology at Burstein Technologies in Irvine, California, invented a technique that transformed biology and won him the Nobel Prize for Chemistry in 1993: the polymerase chain reaction (PCR) technique. Within hours, researchers could create literally millions of copies of short pieces of DNA in which to detect mutations consisting of a single nucleotide change. The technique has revolutionized the diagnosis of cancer and infectious diseases, as well as the field of forensics – in which murderers can be convicted and the wrongly accused exonerated.

What do you find to be the most impressive uses of PCR?

PCR is involved in almost any kind of DNA application. I'm convinced that DNA is one of the most informative molecules that biologists have to work with. Obviously it would be nice to know about all of the proteins and the carbohydrates and lipids too, but it seems from what we know so far that all of the attributes of living systems at some point can be reduced to their DNA sequence. Therefore the study of life itself … can boil down to the study of DNA.

Has anyone come up to you and said, 'PCR changed my life', or that it has put someone in jail?

More important than people who were put in jail are the people that have got out of jail.

Kary Mullis. [Fergus Greer]

1977

Fred Sanger completes the first sequencing of an entire genome – of the bacterial virus, or phage, φ-x174, all 5375 nucleotides.

1977

Phil Sharp, Richard Roberts, Louise Chow and Pierre Chambon discover that the genes of higher organisms and viruses are often fragmented into several pieces (exons) along the chromosome, separated by 'non-coding' regions (introns). A mRNA 'transcript' is then edited by a process called splicing, to produce a 'mature' mRNA that codes directly for protein.

I've gotten a lot of letters where somebody says 'Thank you for inventing this, I've spent ten years on death row, or whatever, and suddenly they looked at the underpants and found out I wasn't the rapist', – that kind of stuff.

That's actually happened?
Yeah, there was some guy who was accused of rape and killing a little girl. The guy was a really colourful muskrat hunter, a trapper. He was being released by Barry Scheck, a lawyer with the Innocence Project. They reopen cases where there's never been a confession, where somebody had been convicted without the use of PCR. It's become a pretty significant thing because they've found out that about 25 per cent of [unconfessed] people, from the DNA point of view, were innocent.

Did you have a strong sense of the potential importance of PCR?
It was pretty obvious to me that [PCR] was going to have important consequences because it solved some fairly important technical problems: abundance and also selectivity. DNA was always the least abundant

"I'm convinced that DNA is one of the most exciting molecules that biologists have to work with."

thing in any experiment – because all DNA seemed physically the same, it was hard to select a particular sequence from amongst the hundreds of thousands of other ones that were found with it in nature.

In order to have that inspiration, is there a parallel between your approach and what Watson and Crick were doing?

I think there's almost an anti-parallel: Watson and Crick knew exactly what they were looking for. They didn't know the form it was going to take – that was their big discovery. I wasn't really looking for a way to amplify DNA, I was looking for a way to look at DNA sequences one at a time. I was also looking for an easy method to find a specific base pair that was different – like a mutation – what they now would call a SNP ('snip', or single nucleotide polymorphism). PCR was an unintended side effect … It turned out that if I did the process over and over again, it would amplify the sequence.

I had the sense to recognize that it was important. It was a really lucky break for me. I don't know why nobody else thought of it. A couple of people, like David Goeddel at Genentech, had gone through the steps without realizing 'I could do this over and over.' I remember when he told me how close he was – he said, 'I could have taken the reagents I'd used in a period of days, just put them all in the same tube, and cycle it – I'd have had my first PCR reaction, but I didn't see that.' That was the one thing I brought to the party – the strong feeling of 'What is the simplest way to approach any particular problem in the laboratory?' – like an innate laziness.

What will you do next?
The next field I'm tooling up for is nanotechnology. It's a much broader field than biotechnology. The buzz words are 'building large molecules from the bottom up', in the same way that biology has always done, and thinking in terms of structures that are even larger than proteins. Biology has done that also. The focus of nanotechnologists is to make structures that are fairly large but to do them in an atomic way – so that you know

TIMELINE
OF MAJOR EVENTS

Machines of the 1980s

1980
Fred Sanger, Paul Berg and Walter Gilbert are awarded the Nobel Prize for Chemistry for 'contributions concerning the determination of base sequences in nucleic acids'. Sanger is one of only four people to have won two Nobel prizes – his first was in 1958 for his work on proteins, particularly insulin. Sanger's method was more amenable to automation than that of Walter Gilbert, and is now the more widely used.

1980s
Aaron Klug and Roger Kornberg unveil the structure of chromatin – the three-dimensional assembly of DNA and protein in the nucleus that enables efficient packing into chromosomes. Helical chains of DNA are normally coiled around proteins called histones, to produce what looks like a string of beads under the electron microscope. These then become 'super-coiled' into an even more condensed form. If stretched out in a line, the DNA of each cell in the body would span 2 metres. Uncoiling provides an opportunity for DNA to be exposed for the production of mRNA and protein.

1983
Barbara McClintock, maize geneticist, is awarded the Nobel Prize for her discovery of 'jumping genes', more technically known as transposons, in which pieces of DNA can move around between chromosomes. This happens in many organisms, from bacteria to humans, and alters gene expression.

> "That was the one thing I brought to the party – the strong feeling of 'What is the simplest way to approach any particular problem in the laboratory?'"

where every atom is, and you can make the same thing over and over again.

Nanotechnology touches on how little things assemble themselves into bigger things, and how those are capable of reproducing themselves. From their emergent properties, what you will get is as exciting as what you get out of life. It's unpredictable. So in a sense you can have evolution, and you can have a very scary element but also a very exciting element. There will be plenty of people who will say 'Hey, don't meddle with that' – anti-futurists who are concerned with crazy scientists inventing something that will destroy the world.

In 1956 you could have known everything you needed to about DNA and its structure, and in 1983 you could have known everything there was to know about DNA enzymology and the kind of stuff that was required for me to invent PCR – there was a finite number of papers in a given area. In nanotechnology it's very hard to know where to start because a lot of things will cross it and impact on it. To me it makes it even more interesting.

Is there a risk that scientists are becoming too specialized and focused to have big breakthroughs?

Breakthroughs have often happened in a specific area but have had a big effect on what people thought about big issues. Big breakthroughs will happen at an accelerating rate because there are so many people there now who work on things that are invisible.

Most chemists think that biology is just a complicated form of chemistry ... And there are plenty of people working in areas like biological chemistry who think that there's something there we'll never understand, that doesn't have anything to do with chemistry, it has to do with God. When I think of something like emergent properties of little systems that are reproducing, it takes the place of that for me. It leaves me to take some wonder in it. But I'm not a religious person – I think that nature itself can provide that. It keeps me being humble without making me worship something.

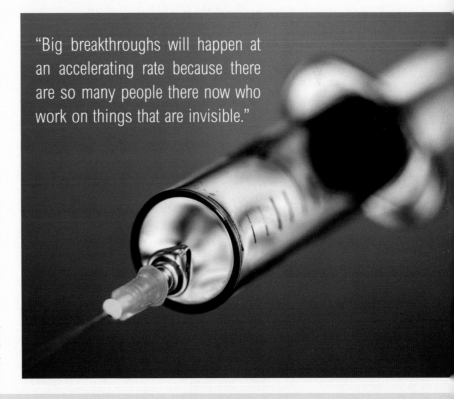

> "Big breakthroughs will happen at an accelerating rate because there are so many people there now who work on things that are invisible."

1984
Alec Jeffries at the University of Leicester invents DNA fingerprinting for identifying individuals – soon to become a key tool in forensics.

1985
Kary Mullis – a biochemist working at the company Cetus in Emeryville, California – invents the polymerase chain reaction (PCR), in which one molecule of DNA can be copied millions of times in a matter of hours. The first application is the diagnosis of a mutation causing sickle-cell anaemia. The technique is now used widely in molecular biology, diagnostics and forensics (see interview with Kary Mullis on page 31). Cetus sells the patent for PCR to the pharmaceutical giant Hoffman-La Roche, Inc. for $300 million.

1986
Leroy Hood, Mike Hunkapiller and Lloyd Smith at the California Institute of Technology invent the first automated gene sequencing machine which reads the sequence of around 250 base pairs a day.

CRAIG VENTER

Craig Venter, founder of The Center for Advancement of Genomics, was still a biochemist at the US government laboratories, the National Institutes of Health (NIH), when he tested one of the first automated sequencing machines in 1986 and was hooked … Here he describes his enthusiasm for sequencing the human genome. In 1998 his company Celera Genomics entered the 'race', and announced the draft sequence jointly with the public consortium in 2001.

Craig Venter. [William Geige]

The mid-1980s was when the first discussion about sequencing the human genome started. I've argued that I'm one of the few, if not the only, biochemist, that decided about the genome project. And my view was driven by pragmatism. I'd just spent ten years of my career trying to find one gene. Here was the genome project offering if it ever got going over the next 20 years, to get all of the human genes … While it's an important milestone, like Crick and Watson sorting out the structure of the double helix, they did not invent the structure of the double helix, they did not create the structure of DNA. Neither did I, or the public effort, create the sequence of the human genome. It has been there throughout our history and we're just the first scientists to read what's already there.

I think getting the first sequence of the genome in 1995, and the sequences of *Drosophila*, human and mouse, are the ultimate expressions of that science that was done 50 years ago. But like that science that was done 50 years ago, these were the earliest baby steps. So I would really like to see a kind of push [forwards] so that during the course of this conversation you could have had your genome sequenced for a few thousand dollars to help plan and interpret your medical future. We can use this to help understand medicine, help understand disease, help understand that race is a social, not a scientific, concept, understand that we're part of the biological continuum and that genetic determinism didn't give us a whole new set of genes over other mammals. Those are the important historical outcomes of sequencing the genome.

TIMELINE OF MAJOR EVENTS

Genomes of the 1990s

1990
Official launch of the Human Genome Project (HGP), an international collaboration of scientists led by James Watson with the ambitious goal of sequencing all 3 billion base pairs of human DNA by 2005. The first phase was to sequence the smaller genome of a multicellular organism, the worm *Caenorhabditis elegans*. The human genome, it is hoped, will provide enormous benefits and new insights into all areas of biology and medicine, including susceptibility to complex genetic disorders such as cancer and heart disease.

1995
First human genome maps produced by Daniel Cohen and Jean Weissenbach at the Genethon laboratory in France, showing the positions of 'markers' – short sequences of DNA scattered across the DNA of all 23 human chromosomes. These will serve as landmarks for securing the locations of larger stretches of DNA now being sequenced.

1995
The first genome sequence of a free-living organism, the bacterium *Haemophilus influenza*, is published by Robert Fleischmann at Johns Hopkins University School of Medicine.

1996
Completion of the first genome sequence for a complex organism, the 'baker's yeast' *Saccharomyces cerevisiae* by an international consortium led by Bart Barrell at the Sanger Centre in Cambridge, and Bob Waterston at Washington University, St Louis.

1997
Ian Wilmut and other researchers at the Roslin Institute in Scotland report the cloning of a sheep named Dolly using the nucleus of a cell from an adult ewe. This historic moment brought the controversial prospects of human cloning a whole lot closer, as previously cloning had been possible only using embryonic cells or cell lines.

The Human Genome Project

What better event to mark the 50th anniversary of the DNA double helix than the completion of the human genome sequence in 2003? This is the reading of the genetic 'blueprint' or 'book of life' containing all the information it takes to build a living human being.

It means having identified and placed in correct order almost every one of the 3 billion chemical 'letters' (the nucleotide bases A, T, G and C) along a 2 metre length of DNA. This DNA is chopped up and tightly bundled into chromosomes, so that two exact copies of the entire sequence exist inside every cell of the body (except red blood cells, which have no nucleus and so no copies, sperm and eggs which have only one complete copy, and certain white blood cells whose genomes undergo radical rearrangements during their development). Now the once-secret code of the body's 'reference manual' is on view to the public on computer databases around the world – thanks to the pioneers who have pushed hard to make it happen.

But like a book written in an unknown language, with only a partial translation available, the interpreters are now busy translating the codes, identifying what genes are there, and what proteins they produce. It holds a vast wealth of information that will keep biologists busy for decades, and promises many new avenues for the diagnosis and treatment of human diseases.

How it all began …

When the idea for the Human Genome Project was first proposed in 1985 by Robert Sinsheimer of the University of California at Santa Cruz, it was met with scepticism. Walter Gilbert of Harvard University was one of those with reservations. Gilbert first 'went in thinking, this is absurdly more complicated than we could ever do', but he and others soon became enthusiasts, and a year later the US Department of Energy announced the 'Human Genome Initiative'. Although the initial goal was to complete by 2005, the development of faster sequencing machines enabled the public consortium to crank up the pace and complete the sequence two years ahead of schedule.

1998
The Maryland-based company Celera announces its intention to sequence the human genome in three years, using the 'shot-gun' method in which the entire stretch of human DNA is shattered in millions of pieces, ready for sequencing, without using the mapping approach of the publicly funded Human Genome Project. Company founder Craig Venter provides one of the five DNA samples used for the project (see interview with Craig Venter).

1998
PE Biosystems develops new fully automated DNA sequencing machines capable of reading 1.5 million base pairs a day (6000 times faster than the 1986 prototype).

1998
Andrew Fire and Craig Mello demonstrated the powerful new role for RNA in blocking gene activity, called RNA interference or RNAi. The experiments launched a raft of studies showing that in contrast to the previously held belief that RNA is simply a dutiful messenger and transporter for protein production, it can also have a significant impact on genes during development.

2001
Celera and the Human Genome Project led by Francis Collins publish at the same time the 'working draft' of the human genome sequence after extensive behind-the-scenes negotiation. The HGP's data are published in *Nature* by Eric Lander, John Sulston and Bob Waterston and Celera's sequence is published in *Science*.

2002
Mouse genome is published in *Nature* by an international consortium of scientists led by Simon Gregory at the renamed Wellcome Trust Sanger Institute in Cambridge.

2003
Claims for the birth of the first human clone by the US cult Clonaid.

2003
Human genome sequence due for completion. More than a hundred bacterial genomes now completed as well as other organisms, including the cress *Arabidopsis thaliana*, the malaria parasite and the mosquito.

Where are they now?

Max Perutz

Max Perutz, the person responsible for bringing Watson and Crick together and giving them the freedom to develop their ideas, died in Cambridge at the age of 88 on 6 February 2002, after a distinguished career uncovering the secrets of proteins through structure. He had already begun to enjoy early success when Crick and Watson joined him at his new MRC unit within the Cavendish Laboratory.

In 1953, the year when Watson and Crick discovered the DNA double helix, Perutz celebrated his own breakthrough: an ingenious method of making atoms more 'visible' in X-ray diffraction photographs by 'isomorphous replacement'. This involved attaching dense atoms of mercury or other heavy metals to molecules within protein crystals. The changes in the X-ray diffraction patterns can be used to identify the exact position of these heavy atoms within

Max Perutz beside a model of haemoglobin, 30 May 1997.
[EPA/PA]

the crystal structure and with these as a first toe-hold the protein's structure can, in theory, be constructed around them. This approach led ultimately to the fine-level structure of the red proteins of blood and muscle, haemoglobin and myoglobin. Perutz's pioneering work earned him the Nobel Prize for Chemistry in 1962, together with his Cambridge colleague John Kendrew. Hugh Huxley wrote in *Nature* (415: 851–2, 2002), 'It seemed almost miraculous that such detailed information was now accessible, and the importance of the work was rapidly recognised.' Kendrew died in Cambridge in August 1997 at the age of 80.

Perutz enjoyed communicating his craft to students and wider audiences. As recorded by art historian Martin Kemp, Perutz once recalled 'the first protein structures revealed wonderful new faces of nature'. Perutz continued to present structures as 'faces' rather than abstract mathematical representations, writing in *Nature* (393: 315, 1998): 'I began giving a course in X-ray crystallography of biologically important molecules for students of biochemistry and other biomedical subjects. In my first lecture I introduced lattice theory, trigonometric functions and Fourier series … but half the students failed to turn up for my second lecture … The following year I replaced my forbidding lecture with a non-mathematical, largely pictorial introduction called "Diffraction Without Tears".'

In his later years, Perutz began pursuing the mystery of Huntington's disease, a neurodegenerative condition involving abnormally long stretches of the amino acid glutamine – called glutamine repeats. These, he proposed, could cause the death of nerve cells by clumping together in large bundles in the cytoplasm. Such bundles have been detected in the brains of diseased patients, but their role has yet to be revealed.

According to Aaron Klug in *Science* (295: 2382–3, 2002), Perutz 'died full of years and of honors, and a happy man, to the admiration and affection of the many who came to visit him in his last few months'.

James Watson

After the excitement of discovering the double helix with Francis Crick, James Watson left for the California Institute of Technology in Pasadena, where he spent two years 'unsuccessfully trying to find the structure of RNA'. In 1955 he returned briefly to the Cavendish Laboratory to work with Francis Crick on virus assembly, before taking a position at Harvard University in 1956. Here, together with his physicist friend Walter Gilbert, Watson investigated the existence of a short-lived go-between, messenger RNA, that converted the DNA code into a meaningful product: protein (see interview with Walter Gilbert, Chapter 3). He also taught biology at Harvard, but was later 'removed' from the course after proclaiming that embryology was a waste of time until DNA became a more accessible tool. 'I thought it was a dead subject and should stay in the deep freeze', he recounted in *Passion for DNA*.

In 1968 he became director of the Cold

James Watson at Harvard University, 1962. [courtesy of the Harvard University Archives]

Getting There by Meryl Taradash, in the grounds of CSHL. [Julie Clayton]

Spring Harbor Laboratory (CSHL) on Long Island, where he attracted new investment and scientists, and forged a new emphasis on molecular biology and cancer. During this time he staunchly defended the use of recombinant DNA technology in the face of 'doomsday' fears that 'we molecular biologists had a genie that we might not be able to contain'.

In 1989 he took on the additional role of Director of the National Center for Human Genome Research at the National Institutes of Health (NIH), and began commuting between Long Island and Washington DC. He campaigned for the launch of the HGP, including an impassioned plea to the US Congress for funds. The politics, however, overwhelmed the science, and Watson left the HGP in 1992 after a clash over the issue of gene patenting, an event he recalls as 'humiliating' (see interview with James Watson, next page).

Nonetheless, his contribution and style are praised by many, including ex-Celera founder Craig Venter:

I credit Jim Watson with playing a critical role in getting the sequencing of the human genome to happen. If he hadn't put his reputation on the line

and come to NIH to get it started I think it's very likely it would have taken years more for it to get going if it ever did. I certainly credit him with helping to stimulate some of my early thinking on this.

John Sulston of the Sanger Institute in Cambridge commented in November 2002:

Jim Watson starts his book with 'I've never seen Francis in modest mood', and the same could be said of Jim. They're both immensely intelligent, Francis particularly. They both are very caring in the way that they actually are interested in the people around them. They both run very successful enterprises by being both tough but both having enormous scientific integrity. Jim Watson became a press man latterly initially because he took charge of the HGP – very flamboyant, very violent, but very much with his heart in the right place.

Watson is now President of CSHL, and has published seven books: *Molecular Biology of the Gene*; *The Double Helix*; *The DNA Story*; *Molecular Biology of the Cell*; *Recombinant DNA: A Short Course*; *Girls, Genes and Gamow*; and *Passion for DNA: Genes, Genomes and Society*.

He and his wife Elizabeth have two sons, Rufus and Duncan.

"Wherever Watson is, in the room or on the printed page, there's tension."

Horace Freeland Judson, *Nature* (413: 775, 25 October 2001)

INTERVIEW WITH JAMES WATSON

In an interview in December 2002, James Watson reflects on the events of 1953 and looks forwards to how genetics may be used in future.

What was the intellectual atmosphere when you first arrived at Cambridge University?

It was the best university in the world. After the war people had a lot of self-confidence because they'd won the war, so in a sense young people felt they were important. Cambridge was important for giving people in some sense the view that they were the chosen people – at different levels.

Pomerat's description of you and Crick after your discovery was of you being like 'Mad hatters at a tea party' – was that true?

I think it was really good [as a description]. I was the first person in Cambridge to be excited about DNA – no one else was. Francis was still dominated by thinking about protein. Perutz and Kendrew ... they thought it [protein structure] would explain life. Max regarded me as a pleasant person to have around but no one expected me to change the world. But the moment Max and John saw it, immediately they knew that DNA was more important than protein in the sense that it was the gene.

What was the chemistry between you and Francis Crick that made you such a successful pair?

It certainly helped that we had the same disdain for anything religious. We believed that the truth comes from experiment and observation and not by revelation, and that some problems were a lot more important than others. If you want to understand life, you've got to study DNA. So we both had a sense of common scientific objectives. In a sense I was the younger brother that he didn't have to compete with. And as the older brother Francis was always very kind and generous.

If you could work now with Francis, what big question would you like to answer?

[For Francis, the question would be] The brain. But my main desire is to stop cancer. Francis has never said for example, 'I want to cure schizophrenia,' so he's never, in that sense, wanted to cure something. But for a long time we've wanted to understand the mind.

Has the whole story been told about the discovery of the double helix?

I really don't think you can say any more. What I read in Brenda [Maddox]'s book – which I think is a really good book – is why didn't we acknowledge more the importance of Rosalind's picture? I can't remember at all. All I can remember is that we didn't think much about Rosalind. Our friend was Maurice and she basically took away his project and didn't want to share it with him. We never expected to get all the credit, we expected to propose a model and later the data came through to prove it. The model was so pretty that we wanted to believe it no matter what the data might say. The big impact of Rosalind's data that she should have known was that there was a helix – the B picture. And she'd had that for a year. We didn't feel that we stole it from her. She had decided to give up DNA and pass on all that she knew to Maurice and that's why he had the picture.

When we got the structure we weren't feeling sorry for Rosalind at all ... But we felt very awkward about Wilkins and of course then we felt very relieved when he got the Nobel Prize because we felt that if he hadn't have made the first step we wouldn't have had the crystalline compound.

In retrospect the structure could have been solved without ever seeing any of Rosalind's photographs. It was easily deducible if you look at Astbury's photograph in the Cold Spring Harbor symposium: it almost looks like it could be a helix. It wasn't as beautiful, but if you looked at it with Chargaff's data and other data that said the bases were hydrogen bonded – basically you had the diameter, you had this 3.4 Ångstrom repeat, and model building would give you the 30 Ångstrom repeat for a single chain. The EM pictures had the diameter.

The 'unit cell' came from the MRC report, but the unit cell I never understood and never used – that is the monoclinic space group which, if correctly interpreted, said that something went up and something went down. So it really should have told Rosalind that it was a two-chain structure, but she never used those data.

James Watson and Francis Crick, 2002. [Kent Schnoeker/Salk]

This is all in retrospect, but the chief thing that influenced me was just seeing the B photograph: it said it was a helix. It confirmed it, at a time when we still hadn't built a model for over a year, and Pauling was proposing a model that we thought was wrong. So it was a new influence. If you looked at the A photo you would say that it's compatible with a helix, but you don't automatically say it's a helix. But if you look at the B photo you'd say 'It's a helix, a perfect helix'.

How long do you think it might have taken Rosalind or Maurice to deduce the structure?
Three to six months, or maybe she wouldn't have [got there] because she wouldn't have talked to a chemist about base pairing – this English reluctance to force yourself upon other people. At first I didn't use the base paired skeleton and I should have. Life is filled with irrational ways of behaving.

If you were to write *The Double Helix* now, would you write it differently?
No, because I was writing it as a novel, I was writing it because I felt that way. At the end of it I say we now realize Rosalind's work was very good. I was not trying to write history, I was trying to write as if it was a 'non-fiction' novel. So I acknowledged Rosalind and how I'd seen her give a seminar and how Maurice described her. People say 'Wouldn't you have written it differ-

ently?' No! Some criticized me for the attention I paid to her glasses, but she did wear glasses; and that I didn't know she was stylish, but I didn't know upper-class dress. I get lots of letters, including from women, who say they enjoyed it [*The Double Helix*]. It was how I felt.

And then you worked with Rosalind later …

When she took up TMV [tobacco mosaic virus] I went and saw her, particularly the year I was in Cambridge. We weren't close but we had a common interest, together with Don Caspar. We weren't working together but we certainly weren't working competitively. She never regarded Francis as having 'done her in', I think the ending of the BBC movie [*Life Story*, 1987] is right. I think she knew that … her reaction was 'I didn't do it – build models.'

Did you talk about that with her?

No – I felt embarrassed. I think everyone would have been. She wanted to, but as a winner you don't … It would have been bad manners … And she was so formidable.

Did you come to see her differently in later years?

Yes, and I was on her side when the Agricultural Research Council didn't give her any money.

Now people write that it [the double helix] was the biggest discovery of the century, but if you look back on those years the majority of people didn't believe it. And it wasn't until Meselson and Stahl in 1958 that it was proved the two chains could separate … It was only after Rosalind's death that there was common acceptance of the structure.

Did Rosalind realize what a big breakthrough it had been?

I'm sure she saw it as the big breakthrough – she was very intelligent. But not as big as now. Now it's the biggest thing of the [twentieth] century, so it's very different. Francis and I thought it was going to be very big, we had a vested interest in it, so we were always trying to encourage our views, but the Cambridge biochemists called it 'the W.C. [water closet] structure'. Even Fred Sanger was quite sceptical. In 1962 he once referred to Francis as 'Oh, the gene man'. And the fact that Francis was not given the professorship in genetics in Cambridge sort of says what Cambridge thought, which was 'We'll wait and see'.

You could say well, why didn't we share and publish communally with Rosalind? But she didn't get the answer. Cambridge thought it was just a big Cambridge find. After the debacle of the wrong model of the polypeptide chains, Cambridge had got the right model. There were two different philosophies for how to do it. I brought the mould down for King's to build models but they never built models. We felt we gave them more than a year to do it. So I don't think Bragg ever felt we behaved badly, and he was a pretty honest man.

Did you discuss with Rosalind how you came to see that data – the B photograph?

No. I just sort of assumed she would have known … It's hard to remember what happened 50 years ago. But I have no evidence that I ever did [discuss it with her]. In my opinion I just wanted to move on and look at new problems. As the winner I didn't want to go beyond it. I didn't feel I'd done anything dishonest. To me it was always something that she should bring up, not me. To me Rosalind was not someone who thought it was good form to cry over past mistakes – there was something new to think about.

Did you feel guilty at all?

No.

Do you think Rosalind's contribution has been adequately recognized now?

Oh, yes. I'm probably indirectly responsible. If I hadn't written a book this story probably never would have been told. I realized it was going to be controversial. After they [the publisher] said they didn't want the title 'Honest Jim', I thought, well, if they don't want that I'll give them 'Base Pairs'.

After your later role in launching the Human Genome Project, has it met your expectations?

Far more. We knew it was going to be good but I didn't expect the mouse would follow so soon. It's wonderful. There are something like 576 protein kinases – it's wonderful.

My first reaction was very negative. I was against it if NIH wasn't in charge because I thought if done incorrectly, that it shouldn't be done and they [the Department of Energy] would have been the wrong people [DOE had set up the Human Genome Initiative in order to debate the idea]. But I'd always wanted a bacterial genome to be done … and most people [of the grant reviewers] wouldn't give me the money. So I would talk about how many genes there are, and recombinant DNA. I became fixated about the bacterial cell. How many proteins are there? How many proteins do you need for life? These were the things that interested me already in the late '50s.

I just saw it [the Human Genome Project] as

a new commodity. It was just a question of could we get the money. We bypassed the Senate and went directly to Congress. Congress gave us the money and then NIH just had to spend it.

And then you left the project?
I was fired by Bernadine Healy [Director of NIH]. It was very unpleasant. I told her 'Our Irish genes don't get along!' Basically it was the patenting. I found it humiliating. There were many people who disliked her, so it wasn't just me, but …

If you had stayed with it, would things have been very different?
No. I really only wanted to do it for four years. Having two jobs was heavy going.

You must be very proud that it's got off the ground.
I'm proud of it but I'm not going to be very happy until we've cured cancer.

What has your focus been at Cold Spring Harbor?
I came here with the thought that we would use the institution to promote not just ourselves but also other people in the world who worked on cancer.

Why cancer?
It's a big problem. I wanted to understand the virus connection. My PhD was on viruses and I was brought up in a world where cells were too complicated and so you started on viruses. I had an idea when I taught a course at Harvard in '54 to '58 that tumour viruses carried genes that turned on DNA replication – I wanted to know if it was right. It was finally proved here.

Where would you like to see genetics being applied in the future?
Everywhere. I don't have a preference. Understanding human nature, I think, is one of the big objectives for this century: to what extent are we really controlled by genes? You just have to talk to Francis Crick's mother to know that he's not a product of his [mother's] upbringing. She was so nice, but they had nothing in common. I don't know what Francis's father was like … His [Francis's] uniqueness comes from the qualities that I found so likeable – how much of it was in his genes? I don't know, but my guess is that it couldn't have been much else.

In Part II of this interview in Chapter 10, James Watson reveals his support for the idea of 'genetic enhancement' – permanently altering the genetic constitution of individuals who have had a 'bad throw of the genetic dice', to the benefit of their future descendants.

Francis Crick

Francis Crick was won over briefly to the study of viruses by his enthusiastic collaborator James Watson, but in 1955 he shifted his interest back again to the question of how the information in DNA is used to make proteins. In 1961, together with Sydney Brenner and George Gamow, Crick 'cracked' the puzzle of the genetic code. In his book *What Mad Pursuit*, Crick likens this to Morse code, with the series of dots and dashes corresponding to the 26 letters of the English alphabet. The genetic code, he said, relates 'the four letter language of the nucleic acids to the twenty letter language of proteins'. They found that nucleic acid triplets each specified the addition of a single amino acid to a growing protein.

In 1962 he became Head of the Division of Molecular Genetics at the newly built MRC Laboratory of Molecular Biology in Cambridge. In the same year he shared the Nobel Prize in Physiology or Medicine with James Watson and Maurice Wilkins 'for their discoveries concerning the molecular structure of nucleic acids and its significance for information transfer in living material'.

But the sunshine beckoned, and in 1976 Crick joined the Salk Institute for Biological Studies in La Jolla, San Diego, California, to begin investigating the other subject that had long held his interest: how the brain works. An agnostic, Crick was particularly fascinated by the nature of consciousness, which he believes boils down to the workings of a bunch of nerve cells rather than anything more 'spiritual'. Taking a theoretical approach, Crick is still pursuing the question of whether consciousness involves a distinct set of nerve cells, or arises from different signalling patterns in nerve cells that carry out other tasks. The best place in which to search for these 'neural correlates of consciousness', according to Crick, is the visual system, which gives us our 'vivid picture of the world and of ourselves'. He collaborates with neuroscientist Christof Koch at Caltech, and in 1994 published a book describing his ideas, *The Astonishing Hypothesis: The Scientific Search for the Soul* (New York: Scribner).

More recently, Koch and Crick outlined their interpretation of case histories of patients with brain damage who have particular deficits in their awareness – who may deny, for example, that they have seen a glass that they have just picked up in their hands. Such cases have led Koch and Crick to speculate about why consciousness has evolved at all in humans, rather than just leaving it all to 'the zombie within':

Online systems are fast, outpacing consciousness. Anecdotal evidence and psychological research emphasize rapid and effortless behaviour that predates consciousness. This is particularly true of the highly practised and ritualized sensorimotor activities that humans love, such as rock-climbing, fencing and dancing. Mastery of these requires a surrendering of the conscious mind, allowing the body to take over.

The hallmarks of a zombie system are stereotypical, limited sensorimotor behaviour, and immediate, rapid action. Its existence raises two questions.

First, why aren't we just big bundles of unconscious zombie agents? Why bother with consciousness, which takes hundreds of milliseconds to set in? It may be because consciousness allows the system to plan future actions, opening up a potentially infinite behavioural repertoire and making explicit memory possible. (Nature, *411: 893, 2001*)

In an interview with the *Los Angeles Times* in November 2002, Crick compared the challenge of defining consciousness with the more straightforward business of DNA structure: 'The double helix was simple because it goes back to the very beginning of life, when things had to be simple. But consciousness is the product of millions of years of evolution.'

He has a son, Michael, from his first marriage; and two daughters, Gabrielle and Jacqueline, with his second wife Odile Crick. Odile drew the illustration of the double helix model for the April 1953 *Nature* paper, and continues to work as an artist.

Francis Crick. [Marc Lieberman/Salk]

Rosalind Franklin. [National Portrait Gallery Picture Library]

Rosalind Franklin

Rosalind Franklin was, by all accounts, far happier after her move to Birkbeck College in London, just before the publication of her April 1953 *Nature* paper describing her X-ray study of DNA. She installed herself in the laboratory of Professor John D. Bernal, where 'she must have found his active support of women students, and his eagerness to promote them, encouraging and endearing', according to her friend Anne Piper (*Trends in Biochemical Sciences*, April 1998: 151–4).

Franklin applied the same talents to investigating the structure of the tobacco mosaic virus (TMV) as she had to DNA. TMV was an impor-

tant topic at the time as it was the first virus to have been isolated and subjected to structural analysis. Franklin used Perutz's new method of isomorphous replacement in her work, to define not only the geometry of the protein sub-units of TMV, but also the long single chain structure of RNA embedded within (the virus's genetic material). Her team included Kenneth Holmes and Aaron Klug, as well as collaboration with Donald Caspar, then at Yale University. In his Nobel speech in 1982, Klug paid tribute to Franklin's influence: 'It was Rosalind Franklin who set me the example of tackling large and difficult problems. Had her life not been cut tragically short, she might well have stood in this place on an earlier occasion.' In $4\frac{1}{2}$ years, she produced 17 papers, and also commenced work on the polio virus, before succumbing to ovarian cancer in April 1958. Klug then took over her role as head of virus structure research.

Watson has expressed a greater appreciation for Franklin during this time, through his own subsequent encounters with her and their shared interest in TMV structure. In September 1954 Franklin visited Watson and his colleagues Leslie Orgel and Sydney Brenner at the Woods Hole Marine Laboratory in Massachussetts. Watson wrote in *Girls, Genes and Gamow*, 'Observing her current friendliness, Leslie took us aside to say that Rosalind's past reputation as difficult to get along with was unbelievable. In his mind, she had been judged most unfairly when working on DNA. Sydney and I had to agree.' Francis and Odile Crick became close friends with Franklin, and she often consulted Francis Crick.

For women scientists today, Franklin's approach to her work serves as a shining example of how to succeed. Caltech crystallographer Pamela Bjorkman said:

> *She's definitely served as a role model for me – there's no question about it – because she kept on, she loved the science, and that's what mattered to her. The glory for the discovery was not important to her. She just wanted results. She was very systematic and careful, and I admired her a lot from what I read – that she persevered in circumstances that must have been extremely difficult to say the least.* Caltech, December 2002 (see also Chapter 5)

Maurice Wilkins

After Rosalind Franklin's departure from King's, Maurice Wilkins, the 'third man of DNA', continued working on the structure of

DNA, and on its packaging into chromatin together with histone proteins. He also pursued the structure of RNA, and of cell membranes. As his colleague Herbert Wilson (his co-author in their April 1953 *Nature* paper) recalls:

> It took Wilkins's group, under his inspiring leadership, a decade to carry out the detailed and meticulous analysis of the X-ray fibre diffraction data. These data firmly established the basic 'correctness' of the double-helix model, although details of the original model had to be modified and refined. If any doubts remained, these were dispelled by the single-crystal studies of oligonucleotides carried out in the early 1980s in the laboratories of Richard Dickerson, Alex Rich, and Olga Kennard. (Trends in Biochemical Sciences, 26, May 2001)

It was only many years after Franklin's death that Wilkins became aware of the existence of Professor John Randall's fateful letter of appointment to Franklin that was to have driven a wedge between them before Franklin had even begun to study DNA (see interview with Maurice Wilkins).

Wilkins continued teaching undergraduates at King's until 2001, when he decided to devote more time to writing his autobiography. He continues, however, to conduct an evening discussion class on 'open dialogue'. In November 2002 Wilkins, now a national scientific hero in New Zealand, was honoured at the age of 86 by the unveiling of a portrait of himself, commissioned by the Royal Society of New Zealand and the New Zealand Portrait Gallery.

[The Royal Society]

Aaron Klug

Rosalind was sharp, quick, incisive, a superb experimenter with strong analytical powers, but she was not as imaginative as somebody like Crick or Pauling. And that may account for one of the reasons that she didn't want to go into model building, because she knew about the 'fiasco' in which Bragg, Kendrew and Perutz had failed to find the alpha helix. Rosalind decided to approach in a more ordered analytical way, rather than inspired model building. Francis describes her as a more cautious person, I would say she's a more systematic person.

Maurice Wilkins and she had very different personalities and they started off on the wrong foot as a result of the [Randall's] letter. It hasn't got anything to do with being a man or a woman. If Rosalind had been a man they wouldn't have got on either. [The idea of misogyny] was a red herring. Rosalind was determined to be treated in her own right, as she was in France. At Birkbeck, in Bernal's lab, they were treated very respectfully. She'd already established herself. [Her] work on the carbons was first class. It was very elegant. When she started on TMV she started by changing the water content – it had no effect. But she took the superb pictures of TMV – Bernal said they were the most beautiful pictures ever shown.

James Watson (left) and Maurice Wilkins (far right) with HRH the Princess Royal at the opening of the Franklin Wilkins building at King's College, London, 2000. [King's College]

INTERVIEW WITH MAURICE WILKINS

January 2003

Maurice Wilkins. [King's College]

Did you ever have a sense that Watson and Crick's model of the double helix would lead to such an explosion in molecular biology?
Within months of the discovery of the double helix people were thinking about how far it would go. It took some time for all the possibilities to be followed up. The human genome sequence was an enormous task but it has now been completed, and things were more rapidly done over the years. The Watson and Crick model wasn't 'the structure, full-stop'. It was a magnificent suggestion but in science you have to move on from brilliant suggestions to finding out the truth, so it had to be followed up. We worked on it at King's for eight to ten years to pin the structure down. The pairing of the bases was a key factor and we were able to test that out with more precise X-ray diffraction studies.

That must have been very satisfying …
I don't think anybody bothered much – they were so carried away by the announcement of the double helix. Most people assumed that the general idea was correct, which was what

> "In the modern world the role of the individual has become much less important than joint interaction between groups of people. I think the main thing is to show this need for co-operation and openness."
>
> Maurice Wilkins

mattered, so I don't know if most workers in the field were terribly excited about what we did. I think there's no doubt it needed to be followed up and put on a sound foundation. By the time of the Nobel Prize we had very much finished that work.

Looking back to the events of 1950 to 1953, what lessons are there for future generations of researchers about the need for collaboration?
When Rosalind Franklin came … there was the administrative muddle. In Paris, she'd done four years' work on coal and similar

things very much under her own steam and established a good reputation. She'd assumed that on coming to King's she would be working on DNA like that. But it wasn't like that at all because she wasn't properly in the picture. It was a very difficult time because she was not the type of scientist who found it easy to collaborate. She was very much happier in her subsequent work at Birkbeck College where she had a greater degree of independence. Scientists in general have some degree of sharing and co-operation, but that's not the way she was. Sometimes people find the need to be left alone to get on with things more in isolation, but that is a relatively ineffective way for people to work.

Brenda's book makes clear that Rosalind was in a very disturbed state when she came to London. For a scientist to have to leave a lab after four years of being very successful she must have found it very difficult. In our lab we had no idea that she had left Paris in a state of distress. If people had known this they might have made bigger efforts to try to lean over backwards to reassure her. She was not a person who was very open about how she felt.

The letter [of appointment from Professor John Randall] was only discovered many years after Rosalind's death. The family handed all her collection of letters to Aaron Klug, and that's when it was discovered. I was assistant director [of the Biophysics Department at King's] and whether Rosalind thought I was anything to do with it, I don't know, I never knew it existed. This was understandably not the way to run science. I was astounded that any such impression had been given to her.

So you were under the impression that Rosalind was coming to work as an assistant?
I don't know about assistant – one of the group. We were looking forward to her joining us.

In a collaborative sense?
Yes, that's the way science works. The basic principle is sharing information. You can only go a certain distance alone. In the

modern world the role of the individual has become much less important than joint interaction between groups of people. I think the main thing is to show this need for co-operation and openness.

Do you remember why you showed Rosalind's 'photograph 51' to James Watson?

Rosalind was leaving any day and so Gosling handed it to me and I took it for granted that it had been recently obtained, and not eight months earlier. The photograph was a very nice one. I deal with these things in some detail in my own account [Wilkins' autobiography is due for publication in the autumn of 2003] but my general policy was to show people in Cambridge new information obtained in the King's laboratory. I had a much improved sperm [DNA] diffraction pattern in 1952 and I wrote to Francis Crick to say how pleased I was. So it was nothing for me to have shown this new pattern to Watson when Rosalind was leaving our lab any day. I didn't think it [the photograph] was out of this world because in any case we'd had the sperm patterns before – showing a series of helical layer lines – but it was good.

Has Rosalind been sufficiently acknowledged now for her role?

There are some things where she did more than is sometimes claimed and in other ways it's very difficult to give her proper credit in view of all the tensions and secrecy. But when the notebooks came out it was much clearer to find out what she'd been doing. We had been completely unaware of some of the important things that she'd been doing.

Rosalind is credited for providing in the MRC report the 20 Ångstrom diameter of DNA ...

Most of the measurements had been around in our lab for some time. Rosalind with Raymond increased the accuracy of these things and I think seeing things down in writing does make a difference to people. My talk in Cambridge in July 1950 said that DNA had a roughly 20 Ångstrom diameter. All these things were quite public, and the 20 Ångstrom diameter was around in Astbury's earlier work. Her [Rosalind Franklin's] pictures looked a bit more flash, but the scientific content ... if I had thought that it would make Jim Watson fly through the air I dare say I might have thought twice about it.

Linus Pauling

Linus Pauling reached the grand age of 93 before he died in August 1994. He may have contributed to his longevity by practicing his latter-day conviction that 'megadoses' of vitamin C would ward off ill health.

Pauling had always been a risk-taker, and is the only person to date ever to receive two unshared Nobel prizes. The first was for chemistry in 1954, in recognition of his work on the nature of the chemical bond, which helped to lay the foundations of molecular biology. The

Linus Pauling speaking at Oregon State University in 1986. [From the Ava Helen and Linus Pauling Papers, Special Collections, Oregon State University.]

second was for peace, in 1962, after his high-profile crusades against the horrors of nuclear weapons – in which he risked personal and scientific freedom during the McCarthy era of the early 1950s in the United States. His activities included numerous public speeches and the presentation of a petition in 1955 to the United Nations containing the signatures of more than 9000 scientists in protest against nuclear tests. He also received many other scientific and humanitarian awards.

In 1964, under pressure from administrators because of his political activities, Pauling left his position at Caltech to set up the Center for the Study of Democratic Institutions in Santa Barbara, at the same time taking up teaching positions at the University of California, San Diego, and Stanford University.

His scientific interests moved on to the potential for using vitamins and minerals as food supplements to prevent disease, particularly high doses of vitamin C for combating influenza, cancer and heart disease. This pursuit led him to set up the Linus Pauling Institute of Science and Medicine, close to Stanford, after retiring from university teaching in 1973. The institute moved to become part of Oregon State University in Corvallis, Oregon, in 1996.

Even in his last few years, Pauling continued to give remarkable lectures on the subjects that interested him most. Pamela Bjorkman remembers:

I heard him talk at Caltech in the early '90s. It was a remarkable talk: he discussed his work that had spanned 70 years, work that started when he was probably an undergraduate at Oregon State University. In the latter part of his life he got interested in some pseudo-crystalline materials that had hexagonal and pentagonal patterns in them. He was in an argument with some other people about whether or not they were really crystalline. In his talk I remember he went back over stuff that was published – it must have been in the '20s – that was related to this problem. And I realized that it was the only talk I was ever going to hear in my life where he pulled up his own research from 70 years ago to support a conclusion that he was making now. And then he did something that he was famous for – he quoted these unit cell constants to four digits past the decimal ... with no notes. (December 2002)

Pauling had 4 children (1 of whom was deceased), 15 grandchildren and 19 great-grandchildren at the last count in 2002.

"Linus Pauling ranks with Galileo, Da Vinci, Shakespeare, Newton, Bach, Faraday, Freud and Einstein as one of the great thinkers and visionaries of the millennium. Truly he was not of this age, but for all time."

Gautam R. Deiraju, *Nature* (408: 407, 2000)

Crystal gazing

When Maurice Wilkins and Rosalind Franklin took their pictures of DNA, the art of using X-rays to peer at tiny molecules was still in its infancy. Now, 50 years later, structural biologists have illuminated every corner of biology with their stunning images, thanks to ever-more sophisticated tools.

According to crystallographer Pamela Bjorkman, the April 1953 papers by Watson and Crick, Wilkins and Franklin and their collaborators gave more than a helpful prod to a fledgling field:

> *I think it made people realize that structure could explain things by simply looking. Sometimes you don't know what you're looking for, and when you solve the structure it can answer questions that you hadn't even thought to ask. Richard Feynman had a great quote, which was if you want to understand something you simply have to look at it, and I think structural biologists believe that and that's why we keep solving structures!*

Wayne Hendrickson, crystallographer at Columbia College of Physicians and Surgeons in New York, agrees:

> *It was so wonderful to have this great, what I call, 'aha' experience – you get the structure and you suddenly understand bunches of stuff about how things work. There's probably no greater example ever than this one in terms of revealing right away the biochemical basis of a major thing – heredity. It put onto a molecular footing a whole battery of experimental work. It was a very fortuitous example.*

Crude preparations

The DNA fibres that Wilkins and Franklin used were crude compared to today's sample preparations. Although fibre work is still useful for some areas – for example naturally fibrous structures such as tendons or to look at the dynamic processes which occur when muscle

Crystals of a complex of DNA and protein. [Bernard O'Hara and Renos Savva/The Wellcome Photo Library]

fibres contract most – structural biologists today obtain the sharpest, most informative images by growing crystals of their material. The use of X-ray diffraction to study crystals is called X-ray crystallography.

Until the 1970s, only the smallest, simplest proteins could be crystallized, but the advent of recombinant DNA technology vastly increased the range of possibilities. Researchers could at last make short synthetic stretches of their favourite DNA, RNA or protein, and in pure form, making them easier to crystallize.

How to grow a crystal

Just as salt forms on the shores of Israel's Dead Sea, a pure solution of either protein or DNA will form crystals upon evaporation in a range of carefully controlled conditions (but the crystals are so tiny they are barely visible to the

Right: Bovine retinal protein rhodopsin. (Palczewski, K. *et al.*, *Science*, 289: 739, 2000) [reproduced with permission of the American Association for the Advancement of Science]

Below right: The gp120 coat protein (red) of HIV virus bound to the cell surface receptor CD4 (yellow) and an antibody fragment (blue). (Kwong, P. D. *et al.*, *Nature*, 393: 648, 1998)

Below: Ribbon diagram showing the structure of a nucleosome core particle: the natural packing arrangement of DNA inside the nucleus, in which the 146 base pair fragment of DNA (turquoise and brown) is wound around 8 histone protein chains (yellow, red, blue and green). (Luger, K. *et al.*, *Nature*, 389: 251, 1997)

naked eye). This occurs when millions of copies of the same molecule line up in exactly the same orientation, in an endlessly repeating pattern as the crystal in all directions. The result is a three-dimensional lattice, held together by physical bonds between the individual molecules. The time required for crystal formation, however, can be anything from a week to years – and some molecules are not amenable at all to crystal formation.

X-ray vision

X-rays are a form of electromagnetic radiation like light or radiowaves but with much shorter wavelengths and greater energy. Consequently they penetrate further into biological material. Wilhelm Roentgen first discovered X-rays by accident in Wurzburg, Germany, in 1895, when he noticed that a screen in his laboratory glowed in the dark after being exposed to a source of radiation – a discharge of electron beams from a gas discharge tube. It would not have been surprising, except that the screen was surrounded by black cardboard, and so was impenetrable by ordinary light. He alerted the world to the phenomenon by taking an X-ray photograph of a hand, in which the bones produced dark shadows. A news release from the London *Standard* newspaper announced the discovery of 'a light which for the purpose of photography will penetrate wood, flesh, cloth, and most other organic substances'.

The majority of X-rays pass through soft tissues such as muscle and fat, where the small atoms of biological molecules – mostly carbon,

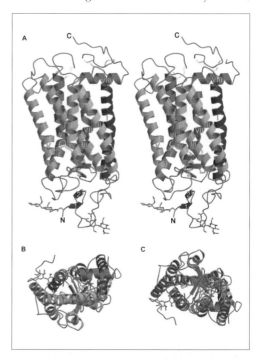

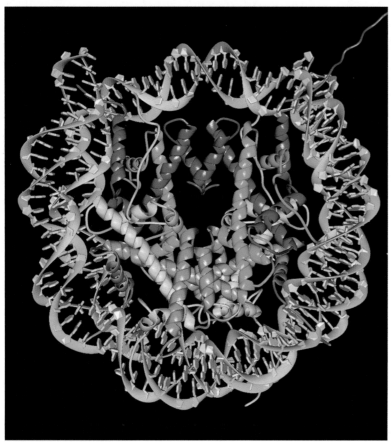

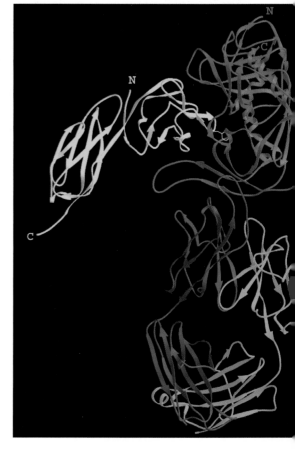

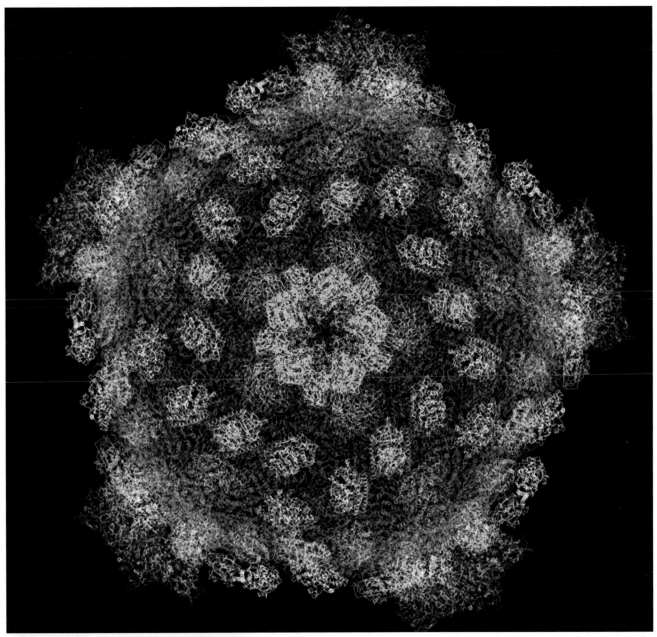

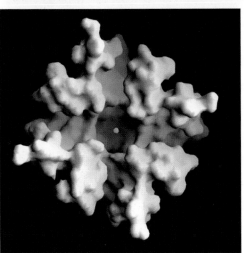

nitrogen, oxygen, phosphorus, sulphur and hydrogen – present little obstacle. In bone, however, X-rays collide with the much larger atoms of calcium, and will either get absorbed or bounce off – rather like two billiard balls colliding. When passing through a crystal, however, its regularity scatters those rays that don't pass straight through in a very specific way called diffraction. Even the light atoms that make up a protein can scatter a few X-rays, so that a diffraction pattern – dictated by the precise arrangements of these atoms – will build up on X-ray film, or on some other sensitive surface placed behind the crystal. The more molecules there are in the same orientation, the sharper and clearer the pattern.

Above: Reovirus core particle. (Reinish, K. M. *et al.*, *Nature*, 404: 960, 2000)

Left: A cell membrane potassium channel. (Jiang, Y. *et al.*, *Nature*, 417: 523, 2002)

Hendrickson described this: 'The crystal is a fabulous amplifying machine that allows you to take very weak signals that would come from individual molecules and accumulate them in a co-ordinated fashion into diffraction patterns which can be measured with very great precision.'

Sir Lawrence Bragg used the simplest crystals available – common table salt, containing molecules with just two atoms each (sodium and chloride) – and found a correspondingly simple diffraction pattern, with a small number of spots widely spaced. In contrast, a crystal of a protein or DNA, containing many thousands of atoms, gives more spots, closer together.

According to Bjorkman, 'If you look at the pattern in all the right places you will see a set of reflections that are obviously the same on one part of the diffraction pattern compared to the other. People used to look at these pictures and figure it out … Nowadays people just run this through a computer program that knows all that.'

Collision course

Structural biologists are now exploiting the gains of high-energy physicists who, in the 1970s, created a new source of brighter X-rays called synchrotron radiation. This involves sending elementary particles whizzing at high speed around a kind of curved atomic racetrack until they crash into one another – the debris being the most informative part of the exercise. As a by-product, however, the acceleration of the particles causes them to release energy in the form of X-rays. Biologists can use filters to select certain wavelengths, in tune with their particular needs.

Synchrotron radiation is faster and brighter than laboratory X-ray sources, revealing more detail more quickly about crystal and fibre structures – just as bright light enables a photographer to take good photographs using a faster shutter speed on a camera. The process can be so quick that researchers can get the data they need within an hour, and solve a structure that same day, according to Bjorkman.

Like scientists turned film-makers, some biologists are now attempting to take several images in sequence to produce 'snap-shots' of proteins as they interact, move and change shape, over time – for instance to understand how enzymes interact with their substrates.

Synchrotron radiation is used for fibre diffraction as well as crystal diffraction – to examine, for example, polymers such as DNA, RNA and collagen, and muscle proteins like myosin and actin. Myosin is amenable to both fibre and crystal structure analysis: the entire molecule is a fibre consisting of a head plus a tail – these aggregate into a coiled-coil structure – but its enzymatic portion (the head) can be removed and crystallized, and its structure has been solved. In order to overcome the expense of using synchrotron facilities, researchers tend to participate in large collaborative projects. These include Daresbury, UK; Grenoble, Switzerland; and Stanford University, USA. There are also projects in Germany, France, Japan, Brazil and the Middle East. Each new machine allows order of magnitude improvements in resolution and speed of results.

Drug design

As the field of HIV/AIDS medicine illustrates, structural biology is now making an impact in the search and design of new drugs. The investigation of the structure of HIV protease, an enzyme which the human immunodeficiency virus uses to reproduce itself, has led directly

Wayne Hendrickson. [Stephen J. Douglas/HHMI/Columbia University]

"For me one of the really exciting things about structural biology is that we are beginning to nail key things. We now know for example how DNA is packaged into chromosomes [wrapped around proteins to form bead-like structures called nucleosomes]. We know how the transcription process initiates … We don't have the whole picture but we have some of the core elements, the very heart of the Central Dogma is there in those pictures [proposed by Crick to say that DNA directs the production of proteins via an RNA intermediate]. And now in the last couple of years we've had [the investigation of] RNA polymerase in yeast. This is a completely marvellous story."

Wayne Hendrickson

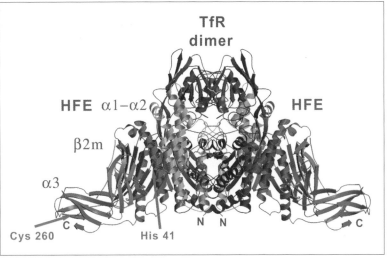

TfR
dimer

HFE α1–α2   HFE

β2m

α3

Cys 260   His 41

Left: At the California Institute of Technology, Pamela Bjorkman has used X-ray crystallography and other structural analysis techniques to investigate molecules used by the immune system to recognize viruses and bacteria, and other proteins with similar three-dimensional shapes that may have evolved from a common ancestral protein.

Above: Hereditary haemochromatosis protein HFE. (*Nature*, 403: 46, 2000)

to the design of compounds that can block the enzyme, many of which are now approved drugs for the treatment of HIV infection. Hendrickson said: 'Structural biology contributed to the development of compounds [that block HIV protease]. We saved lives a little earlier because of structural biology. We can save more lives, and be there earlier still in future, knowing what we know now with new technology. We have a lot more prowess now.'

Besides admiring the work of Rosalind Franklin, Bjorkman also recognizes the work of another great influence, Dorothy Crowfoot Hodgkin (1910–94). When working with Bernal in the 1930s Hodgkin first demonstrated that protein crystals could produce diffraction patterns. She solved the structure of the first biological molecule, vitamin B12, for which she won the Nobel Prize for Chemistry in 1964. She was only the second woman (the first being Marie Curie) to do so – and remains the only British woman to have won the prize. She trained generations of women in the field and encouraged countless young women to enter a scientific career. In Bjorkman's view, 'at a time when the physical sciences were dominated by men, X-ray crystallography could have been totally dominated by men. And this one woman made a tremendous difference.'

Genomes galore

Background image: Normal human karyotype. [courtesy of Joy Delhanty, UCL]

Celebration time

Watson and Crick certainly attracted attention in April 1953 for discovering the double helix, but their paper was just over one page in length and entirely theoretical. Now, in 2003, barely a month goes by without the announcement of yet another genome sequence, adding to the enormous treasure trove of information that Watson and Crick could only have dreamed of 50 years ago. A symbolic gesture to this 50th anniversary, the human genome sequence – all 3.2 billion letters of DNA code – is now virtually complete. It has been a mammoth task, involving 20 laboratories and hundreds of researchers around the world. It is a magnificent achievement in just 12 years. And yet the real work, say some, is only just beginning.

When a draft of the human genome sequence was unveiled in *Science* and *Nature* in February 2001, David Baltimore conveyed the excitement felt by many:

> *I've seen a lot of exciting biology emerge over the past 40 years. But chills still ran down my spine when I first read the paper that describes the outline of our genome ... Not that many questions are definitively answered – for conceptual impact, it does not hold a candle to Watson and Crick's 1953 paper describing the structure of DNA. Nonetheless, it is a seminal paper, launching the era of post-genomic science.* (Nature News and Views, *409: 814, 2001*)

"I've seen a lot of exciting biology emerge over the past 40 years. But chills still ran down my spine when I first read the paper that describes the outline of our genome ..."

David Baltimore, *Nature*, 2001

The draft revealed over 90 per cent of the sequence, with each base sequenced to at least 99.9 per cent accuracy, and about one-third of this sequenced to 99.99 per cent accuracy – in other words, repeated sequencing efforts showed an error rate of less than one error per 10,000 bases.

Surprise

The draft sequence brought an enormous surprise: the finding that humans are built from a mere 31 000 or so genes rather than the previous guess of 50 000 or more (some estimates went up to 150 000). In September 2002 a further reduced estimate to between 22 000 and 27 000 was proposed. An average gene occupies around 3000 bases, although the largest known so far – coding for the muscle protein dystrophin – has 2.4 million bases.

Perhaps even more curious is that genes occupy only a small part of our DNA – just over a quarter – or, in other words, only around 54 centimetres of the 2 metres of DNA packed into each cell. In between are vast desert-like stretches of DNA that code for nothing. Much of this appears to be 'junk' left over from our evolutionary past. But like discovering an oasis, scientists have also found regions within the desert that are important to life today: bits of DNA that help to maintain the shape of chromosomes, and others which act like a thermostat, receiving signals from proteins to switch genes on and off. The largest chromosome – chromosome 1 – has the most genes (around 2968), while the smallest – Y – chromosome has the fewest (approximately 231).

What makes humans so complex?

The lower than expected number of human genes makes even more vexing the question of what makes us so complex compared to other animals – in our body design, our larger brains, our behaviour and language, and so forth. The fruit fly *Drosophila melanogaster*, for example, has 13 000 genes; and the mustard weed *Arabidopsis thaliana* has 26 000 genes – a number not far off from humans! And the mouse has roughly the same number as a human, differing only in that about three hundred genes code for proteins that appear to be unique for the two species (see below).

So other phenomena besides gene number are likely to contribute to the greater complexity of humans. One appears to be the large number of proteins called 'transcription factors' that switch genes on and off. Our development over nine months from a single fertilized egg into a complex body of around 10^{14} cells involves a multitude of successive genetic events in different cells controlled by these transcription factors, usually in response to signals from other cells.

Mix 'n' match

In addition, around 60 per cent of our genes code for two or more proteins, compared to only 22 per cent of genes in the roundworm *Caenorhabditis elegans*. This results from a process analogous to how scenes of a film end up on the cutting-room floor. The initial 'immature' mRNA has an imprint of all the exons (protein-coding portions) and introns of the gene on the DNA template; it is cut up and bits are discarded before other pieces are rejoined in different combinations ('alternative splicing'). The result is a variety of 'mature' mRNAs with varying exon content, which give rise to proteins of different shape and function.

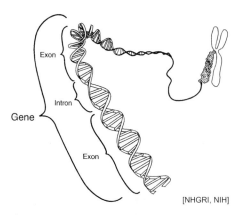

[NHGRI, NIH]

Human genome cover (*Nature*, 409: 754–964, 2001). [Image by Eric Lander, created by Runaway Technology Inc. using PhotoMosaic by Robert Silvers from original artwork by Darryl Leja, courtesy of the Whitehead Institute for Biomedical Research. Gregor Mendel, James Watson and Francis Crick are in the crowd.]

Ancient relics

The human genome sequence also provides a window onto our genetic history. There are extra copies of many genes, for example, due to 'gene duplication', which may help to ensure that certain vital functions are not lost through either mutation or loss of DNA during cell division and gamete formation. About 8 per cent of the genome also consists of virus-like pieces of DNA, possibly the fossils of long-dead organisms that infected our ancestors.

Maynard Olson [B.H. Olson]

MAYNARD OLSON is Professor of Genome Sciences and of Medicine at the University of Washington, Seattle. Initially a 'sceptical chemist', he was impressed by James Watson's assertion in the 1970s that DNA contained a finite amount of information in biology: 'To me it was a tremendously exciting idea that a phenomenon as complex as biology had this kind of bound on it – this finite information content of the genome. Really from that point forward I had this clear idea that the path forward in biology was to define the genome and build a secure foundation.'

Sea of data

Walter Gilbert (see Chapter 3) predicts that it may take the next hundred years to wade through and understand fully the sea of data that is now coming from the human and other genome projects. The task also requires a tremendous amount of computing power for comparing and predicting possible functions of proteins, which in itself has given rise to the new field of bioinformatics. Although this may sound like a daunting task, the comforting notion – according to Maynard Olson – is that ultimately the amount of information that scientists need to understand is finite.

Before genome analysis took off, there was, according to Olson, a 'common biologists' complaint' that biology was impossible to unravel; it was far too complex and involved infinite numbers of interactions between molecules. Cue the genome: a finite, neatly packaged reference guide with no endless stream of supplements or hyper-links – just the genome itself. 'Until we could acquire these genome sequences, doing biol-

ogy had this element of sinking into quicksand, and you had no assurance that there was any solid bottom. And that's what the genome is – it should be immensely welcome to all biologists.'

Olson also believes that the way to exploit the genome is to have very practically based goals, such as curing particular diseases, rather than endlessly cataloguing the interactions of proteins.

Others, however, maintain that a broader, less goal-directed enquiry remains a high priority. The Sanger Institute's John Sulston ranks his priority:

I want to know how life works. I want to get a real sense of how the genes are read out from the DNA in the context of course, of the fertilized egg. DNA, of course, does not create life, DNA maintains life in each generation because it only does it in the context of the life that's gone before. I don't think I shall understand how it works by knowing how a particular bug causes a particular disease, although that may be a part of the story.

Our family and other animals
'Man's new best friend'

The genome sequences of many other organisms are appearing in the literature thick and fast, including the house mouse, *Mus musculis*. Published in December 2002, the public version of the mouse genome sequence, together with the privately owned version completed earlier by Celera, promises to be particularly useful in identifying genes shared with humans that are involved in disease, which could therefore serve as targets for testing new therapies.

'The mouse is an excellent surrogate for exploring human biology', wrote Allan Bradley, who took over from Sulston in November 2000 as Director of the Wellcome Trust Sanger Institute in Cambridge. Other collaborators were based in Spain, Switzerland, Germany, the USA, Canada, Australia and Japan.

Despite mouse and human ancestors having diverged more than 75 million years ago, so far there appear to be as few as 300 genes that distinguish us from our rodent relation (1 per cent of the total). Many more genes in the mouse genome appear to be involved in smell – which is perhaps not surprising given the importance of smell for finding food and choosing a mate to this organism. Furthermore, many more mouse than human genes are involved in eliminating poisons.

Researchers at the University of Geneva Medical School in Switzerland and the Telethon Institute of Genetics and Medicine in Naples, Italy, have created an 'atlas' of genes in the mouse that have human counterparts, showing where they are switched on during embryonic and foetal life. Focusing on the mouse counterparts of human chromosome 21, they have discovered the genes to be active in the developing brain, heart, face and fingers. This corresponds well to the parts of the body affected in Down's syndrome – in which an extra copy (trisomy) of chromosome 21 causes a variety of abnormalities – and will be useful in understanding the condition.

Closest cousins

The rat genome is also now nearing completion but, not content with rodents, scientists at the RIKEN Genomic Sciences Center in Yokohama, Japan, have begun sequencing the genome of our closest relative, the chimpanzee. Because humans and chimpanzees share nearly 99 per cent of their genomes, comparing the genes of each organism may provide the best route yet to discovering what makes us distinctly human.

Each new addition to the genome league creates unprecedented opportunities to understand evolution. The next candidates include the honey bee, the chicken, the cow and possibly even the kangaroo.

Of micro-organisms, we now have the sequences for more than 70 different bacte-

Alice's Adventures in Wonderland, Lewis Carroll, Macmillan.

ria, including the agent of bubonic plague, *Yersinia pestis*, and the lung pathogen *Mycobacterium tuberculosis*. Scientists hope this will assist in understanding more about exactly how they cause disease, and how we might thwart their attacks in future.

More than just trophies

These many genome projects are creating valuable tools for making comparisons between organisms. It may appear paradoxical, but sequencing the genomes of so many organisms makes analysing the data easier rather than more difficult, because genes can be identified by their similarity with those of other creatures. The longer ago in evolutionary history two animals or plants diverged, the more likely it is that any genes they still share today will be involved in important life-sustaining processes, such as energy metabolism. Thus a 'similarity' search across many genomes is one of biologists' most important tools for probing DNA sequences.

John Sulston [The Wellcome Photo Library]

JOHN SULSTON shared a Nobel prize with Sydney Brenner in 2002 for his pioneering work defining the developmental biology of the nematode worm *C. elegans*. This in turn was chosen as the first organism with which to pilot the tools and methods of the genome project, as a 'demonstration model' of how successful the genome project could be. He has commented:

He [James Watson] must have been aware that they [the genomes of simpler organisms] would be important in their own right and important in finding out how the human works. I guess none of us were aware how close the ties would be at that point. Until we started large-scale sequencing I don't think we knew the extent of the unity of life at the DNA level. And that of course is what's really powered genomics. It's the comparisons, and the fact that people go into databases and find hits with quite distant organisms.

Tracing human origins

Out of Africa

Out of Africa we may have come, but what happened next? Archaeologists have provided ample evidence that our modern human ancestors, *Homo sapiens*, spread from Africa to the rest of the world between about 200 000 and 50 000 years ago. For many years, however, they have debated whether these explorers completely replaced the local populations of early humans, including the Neanderthals – whose ancestors had left Africa around 1.5 million years ago – or whether they interbred for a while before gradually taking over.

At the more recent end of the time scale, we know that human populations have continued – and still do – to move around the globe, whether driven by war, famine and plague or simply curiosity and adventure, to search for new and richer homelands. Yet ancestry plays an important part in the cultural identity of people, wherever they settle.

Story makers

In many cultures, stories of ancient conquests and travels have been passed down over generations in the oral tradition, while others have more recently recorded their histories faithfully in writing. But oral and written records, and archaeological evidence, cannot paint the whole picture – some events will have been forgotten or skewed by personal bias, while people's homes and belongings are mostly destroyed over time. The most enduring and potentially accurate record of our ancestry lies not in what's around us, but in what's within us – etched into our DNA.

The molecular biology revolution has pro-

vided an entirely new set of tools with which to re-examine old hypotheses and create new ones, and for helping to affirm cultural identity – for example among Jews and Native Americans. In turn, the study of different populations around the globe is providing clues about the origins of human disease that will help us move into the future.

On the move

There are two theories about what happened when groups of *Homo sapiens* began to fan out from Africa. Most archaeological and genetic evidence has pointed to the formerly popular 'replacement' theory, by which the new arrivals evolved separately and gradually replaced their early human neighbours; but more recently new data have rocked the boat in suggesting that the African emigrants instead interbred with the local populations.

Tracking the movement of people between 10 000 and 30 000 years ago across Europe and Asia, the settlement of Oceania, and the migration from Asia into North and then South America, across what is now the Behring Straits are all studies that are clarifying our evolutionary and migratory history. Native Americans, according to these types of study of the genetic record, are the descendants of a small proportion of the peoples inhabiting Asia more than around 30 000 years ago, having gone through a 'bottleneck' before then spreading out across a new continent.

Male and female lines of inquiry

There are many ways to examine genetic similarity between individuals, and between popula-

"We have history of kings and queens, but with genetics you can have the history of the people."

Mark Thomas, University College London

tion groups. One is to look specifically at the portion of DNA that passes only from father to son, via the Y chromosome, or that goes from mother to daughter, via mitochondria. The latter are the energy powerhouses that exist in the cytoplasm of all cells, in both males and females, membrane-bound bags of enzymes that also contain a handful of genes coding for some of these enzymes. Because only the female egg provides cytoplasm to the embryo (the sperm simply injects its nucleus-filled 'head' into the egg), females only get to pass their mitochondrial DNA onto their offspring.

Alternatively, scientists look at the common portions of DNA shared between both males and females, located across the entire genome, on the autosomes (all chromosomes except the X and Y 'sex' chromosomes).

Fighting fit

The early studies looked at blood groups, and later at other types of protein or the genes encoding them, that showed sufficient variation from one individual to another to provide information about genetic relatedness. One example is an enormous cluster or 'family' of genes called the major histocompatibility complex, or MHC. These code for a whole range of proteins that trigger white blood cells into action against viruses, bacteria and parasites. (The MHC got its name from the role it was first noted for – the 'recognition' of transplanted tissues such as kidney, heart or lung as foreign, leading to rejection if immunosuppressant drugs are not given.)

By studying variations in the MHC genes of different populations, immunologists have discovered how infectious disease in different parts of the world may have become a driving force for evolution. As they encounter new climates and new pathogens (many transmitted by insect bites), only those individuals possessing the right genetic make-up will defeat the pathogens – or at least manage to live with them. After Asians entered North America and moved south over generations, pathogens will have taken their toll, 'weeding out' those with an ill-equipped immune system. The further south Native Americans live today, the more novel variants of MHC genes geneticists find, indicating that these may have assisted the survival of their forebears in reaching that part of the world.

Magic markers

Despite their usefulness, blood groups and MHC genes still present too narrow a constellation of genes to carry out some of the more detailed population tracking exercises that geneticists would like to do. One of the latest tools, following hot on the heels of the Human Genome Project (HGP), is the discovery of alternative spellings of the words in the DNA code called SNPs ('snips') or single nucleotide polymorphisms. Forming a kind of genetic identity tag, these have arisen through the substitution of one nucleotide base pair for another in the DNA sequence, at intervals along the human genome, through errors in the DNA copying mechanism during the formation of eggs and sperm.

Another type of change that geneticists can detect is in the 'microsatellite markers', short pieces of DNA (3 or 4 bases, such as GATA) which over eons of time can be repeated in

tandem along the DNA sequence, maybe 10 or 12 or more times. The repeat number goes up or down over time (although not in every generation), and so looking at the repeat numbers of several different microsatellite markers that sit fairly close together on a chromosome will give an idea of the passage of time since the marker first appeared. Likewise, 'minisatellite markers' also exist, consisting of repeated stretches of between 10 and 60 bases, which change more frequently in repeat number due to the process of 'recombination', in which two chromosomes of a pair swap bits of their DNA sequences.

The good thing about microsatellites is that they are extremely variable. So if you take two individuals picked from a population at random, there's a very high probability that they would have a different number of repeat units at the same microsatellites.

Taken together, SNPs and satellite markers can reveal not only the presence of rare

Genetic analysis of the inhabitants of Antioquia, a mountainous region of northwest Columbia, reveals their ancestral mothers to be Native Americans, while their ancestral fathers were invading Spanish colonials. A higher than expected number of Antioquian families also appear to be susceptible to juvenile diabetes and psychiatric disorders such as manic depression, indicating a 'founder effect'. [courtesy of Andres Ruiz-Linares, UCL]

mutations in someone's DNA – identity tags that they may or may not share with their relatives – but also the time since that mutation is likely to have first appeared in the population. One study, for example, has dated the common ancestor of 16 ethnically diverse men as having lived at around 188 000 years ago, using markers on their Y chromosome.

The lost tribes of Israel

Satellite markers have been particularly useful not only for considering the traditional questions of how and when modern humans evolved, but also for looking at our more recent cultural history.

In today's Zimbabwe and South Africa, for example, there are around 50 000 Bantu-speaking people of the Lemba tribe, who look and sound very similar to other black Africans in the same region but have some customs that are rather at odds with their fellows. These include, in some clans, the refusal to eat pork and the

Lemba tribe.

[Tudor Parfitt, SOAS]

practice of circumcision – traditions normally, although not exclusively, associated with Judaism. For centuries the Lemba tribe told the story that their forefathers were descended from Jewish traders in the Yemen who 'came from the north by boat'. This has been interpreted to suggest that they may be one of the mythical 'lost tribes of Israel', descended from the high priests of the court of King David. But only in recent years has scientific evidence appeared to support the claim.

Mark Thomas, a geneticist at University College London (UCL), and his team have used microsatellite markers on the Y chromosomes of Lemba men to examine their claim. In one clan in particular, they found a very close match with a version of the Y chromosome shared by Ashkenazi priests (a European Jewish priesthood that claims to be the direct father-to-son descendants of Aaron, the brother of Moses, as described in the Old Testament). The date of divergence from the ancestral Y chromosome was, for the Lemba tribe, around 3000 to 5000 years ago, which fits well with the estimated origins of the chromosome of the Ashkenazi 'Cohen' priests. In contrast, the team has not found supportive evidence for a similar claim from a tribe in Nigeria.

'Certainly in the case of the Lemba there is persuasive support for their claims, and in other populations,' said Thomas, who views the human genome as 'a history book or an archaeological resource'.

The studies have impact beyond the world of academia, particularly where pride is placed in cultural identity. Tudor Parfitt from the School of African and Oriental Studies in London, who has worked with the Lemba tribe for years, was rewarded for the findings by being made an honorary tribe member. He said: 'It seems to have made them serious candidates for membership of the Jewish people as far as a lot of liberal Jews outside of Africa are concerned. It seems also to have … increased their own sense of being Jewish.' Parfitt also has evidence for similar links between a group of people in India and Middle Eastern Jews.

Closer to home

When geneticists set off to investigate claims of ancestry they often travel to some of the world's most exotic and beautiful places, encountering cultures far different from their own. But there may also be genetic treasure-troves closer to home. 'I think it's fascinating that people go to all the corners of the world to collect samples but there are some really interesting things going on right here under our noses,' said Erika

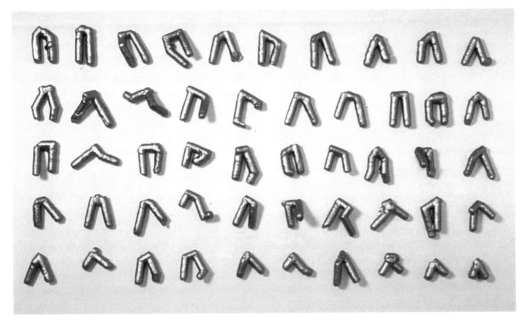

Zoosemiotics (Fish), 1993, by Suzanne Anker, hydrocal, metallic pigment, 36 x 65 x 4 in. [courtesy: Universal Concepts Unlimited, NYC]

Hagelberg of the University of Oslo in Norway.

Hagelberg was the first to discover DNA in ancient bone. During the late 1980s and early 1990s she worked at the Universities of Oxford and Cambridge on the first uses of DNA typing in forensics, including the identification of the bones of Nazi doctor Joseph Mengele. She unearthed some peculiar findings in the Somerset village of Bleadon, in England, which appears on the surface to be the epitome of Englishness. She has reported: 'There's a trace of something very weird and wonderful and old in Bleadon.'

Hagelberg compared the DNA of an Iron Age skeleton in Bleadon with that of 48 current residents of the village. She looked at mitochondrial DNA (passed from mother to daughter). Besides the expected similarities with other people of European descent, there was also evidence of DNA of a more 'exotic' origin, 'very rare mitochondrial sequences' not catalogued in scientific databases as typically European. 'I was intrigued by that,' said Hagelberg.

Although she does not yet know the origin of the stranger ancestors, the findings reinforce for Hagelberg the notion that everyone's DNA has a tale to tell about migration and change of culture.

Hagelberg's work has also involved the analysis of DNA extracted from archaeological bones found on Easter Island in the Pacific, where enormous statues stand as a ghostly sign of a lost culture on the island. She found convincing evidence that the islanders had originated from Polynesia, rather than the alternatively mooted South America. Hagelberg is the first to admit,

however, that such conclusions are never absolute. She cautioned: 'Sometimes people say, "We've solved the Pacific", but we haven't. We've got something that fits in maybe a little bit more accurately than what we thought.'

Hagelberg published recently in the journal *Current Biology* her conclusions about a group of people called the Andamanese, who occupy the Andaman Islands in the Bay of Bengal, off the southern coast of India. They bear a physical resemblance to African pygmies, and their culture and language (only studied in the past

"It's fascinating that people go to all the corners of the world to collect samples but there are some really interesting things going on right here under our noses."

Erika Hagelberg, University of Oslo, Norway

hundred years following British invasion of the island) seemed quite distinct from other people in South-East Asia. Their origins were a mystery. Hagelberg used mitochondrial DNA and Y chromosome samples from people living on the island today, and compared these with samples extracted from Andamanese human hair samples that were nearly a hundred years old. She found that the Andamanese split from other Asian stock probably tens of thousands of years ago. There were distinct genetic differences from other Asian and African populations, and evidence that the Andamanese are descendants of the Paleolithic people (who left Africa during the last Ice Age) who were once more widespread in Asia than today.

What happened to the menfolk?

Meanwhile, on British soil, a debate has raged over what happened when the Anglo-Saxons of Europe invaded between 400 to 800 AD. Was there a mass migration to replace large numbers of the indigenous people, or was there a subtle take-over bid by a smaller ruling elite, who imposed their lifestyle and culture on their new 'hosts'? Archaeologists largely rejected the first idea in the 1960s and 1970s, but Thomas and his UCL team have discovered that the former notion does appear to be closer to the truth.

They took mouth swabs from 313 men living across Britain, from East Anglia to North Wales, extracted the DNA and studied SNPs and microsatellite markers on their Y chromosomes, and compared these with an additional 177 samples collected from Norway and Friesland in The Netherlands. The team found evidence for a mass migration from north-western Europe, at least of men, into central England. This proba-

bly replaced at least half, if not all, the men that were present. Thomas said: 'so now between 50 and 100 per cent of the male line genetic component is descended from invaders rather than from indigenous people'. He concluded, 'it seems as though the Welsh hills have been a bigger barrier to gene flow than the North Sea'.

> ## "If the science is done well, then people don't tend to take offence."
> Mark Thomas

(Gene flow is the mixing of genes through interbreeding between different populations, so that the descendants share genes from both sets of ancestors.)

What exactly happened, however, is open to conjecture. Possibly the invaders intermarried and prevented indigenous men from fathering children. Thomas conjectured: 'They could have been driven out, killed, out-competed, a caste system created – there are lots of different ways that they could have driven out the native Y-chromosomes.'

Danger of misinterpretation

Since the early days of eugenics at the beginning of the 20th century (attempts to use science to justify ethnic divisions), geneticists have always run the risk of having their work misinterpreted as providing a genetic basis for social divisions. But modern genetics, particularly the sequencing of the human genome, has provided the clearest evidence yet that any concepts of 'race' are genetically unjustified. A white person in Europe or the USA, for example, may have more in common, genetically speaking, with a black person in Africa than with their white neighbours living next door. The genes responsible for skin colour and facial appearance, for example, carry only a minute fraction of the variation that exists as a whole between individuals. Furthermore, everyone on the planet shares 99.9 per cent of their DNA with everyone else, with the total variation between them existing in just 0.1 per cent of their DNA.

Despite these striking insights, the new information that genetics brings to understanding the similarities and differences between populations needs to remain in its proper context, so that it does not get misinterpreted as supporting political or cultural ideology, according to Thomas.

He said: 'If the science is done well, then people don't tend to take offence. People take offence when the science, or the way that the

Zimbabwe. [Julie Clayton]

evidence is presented, is done badly.' He referred to a recent study which had to be withdrawn after a 'very unfriendly response' because it had made political points rather than sticking to the facts. A scientific paper, he said, 'doesn't really seem to be the right forum for making political comments'.

A recent study looked at the DNA of over a thousand individuals living in different world regions – Africa, Eurasia (Europe, the Middle East, Central and South Asia), East Asia, Oceania and the Americas – whose families have been there for generations. The results show that of the tiny amount of variation in DNA, most of it (more than 93 to 95 per cent) is among individuals of the same population in the same geographical region. No more than about 5 per cent of the variations were due to differences between major population groups. And the greatest diversity of all is among people living in Africa – which is not surprising, given

that it was just a few of their ancestors who left the continent to populate the rest of the world, and therefore a very small genetic sample.

It is worth remembering, however, that genetics by itself does not replace the need for archaeological and linguistic evidence. In a recent essay in *Nature* (20 June 2002), Austin Hughes of the University of South Carolina, for example, argued that care should be taken over trying to estimate the dates when a population split in two, because two different versions of a gene may have appeared in an ancestral population long before one group departed, taking with it one of the gene variants. The common ancestral gene may therefore be more ancient than the time at which the two populations separated. 'It is true that genetic markers – unlike ancient chroniclers – do not lie. But their interpretation raises many a thorny problem and can be as perilous as attempts to decipher ancient inscriptions in an unknown tongue.'

"It is true that genetic markers – unlike ancient chroniclers – do not lie. But their interpretation raises many a thorny problem and can be as perilous as attempts to decipher ancient inscriptions in an unknown tongue."

Austin Hughes

Gene detectives

Family ties

In the remote village of San Luis by the shore of Lake Maracaibo in Venezuela lives a community unusually heavily afflicted with an inherited nerve-wasting condition called Huntington's disease. People with this disease develop involuntary limb movements when in their 30s or 40s, followed by paralysis. Bizarrely, the disease gets worse over successive generations.

The village has become a classic example of the power of modern genetics. In 1978 geneticist Nancy Wexler from the US National Institutes of Health began compiling a chart of all affected individuals and their families – known as a pedigree – which added up to around 10 000 people, all descended from one woman, Maria Conception, who died in the 1800s. Wexler took around 2000 DNA samples with which to perform 'gene mapping' experiments, painstakingly narrowing down the gene responsible by discovering other known genes that seemed to be inherited together with the Huntington's disease gene, and must therefore lie close to it on one of the chromosomes. The strategy paid off, and in 1993 Wexler finally found the gene on the tip of chromosome 4. Dubbed Huntingtin, the abnormal gene produces a protein with extra long repeat stretches of the amino acid glutamine. In affected individuals, the lengthier pieces cause the protein to clump inside nerve cells, eventually killing them.

Huntington's disease, it turns out, is a 'dominant' trait, in which the mutated gene dominates over the normal gene, and an affected person has a 50 per cent chance of passing the gene to their child (depending on whether the sperm or egg contains the mutated gene). It also illustrates the phenomenon of a 'founder effect', in which a gene mutation occurring in a single individual can be passed on to many descendants. Huntington's disease is prevalent in white South Africans; a number of Afrikaners, for example, inherited the condition from a single Dutch settler of the 1650s.

The commonest lethal genetic disorder is cystic fibrosis (CF), the lung disease characterized by sticky mucous and inability to clear bacterial infections which often kill sufferers by the age of 30. In contrast to Huntington's disease, CF is caused by a 'recessive' gene mutation, two copies of which have to be inherited, one from each parent, if a child is to become affected. Anyone with just a single copy of the abnormal gene – which codes for a protein involved in the transport of salt across cell membranes – is a 'silent carrier'. Around 1 in 25 white people of European descent is the carrier of an abnormal CF gene – caused by any of up to more than 850 different mutations.

Choosing babies

Since the mid-1980s, gene sleuths have identified over a hundred different single-gene disorders, providing a whole new opportunity for genetic testing to help prospective parents avoid giving birth to affected children. Well-known chromosomal disorders such as Down's syndrome (in which individuals have three

Right: FISH analysis of a human embryonic nucleus reveals normal staining for chromosomes tested. [courtesy of Joy Delhanty, UCL]

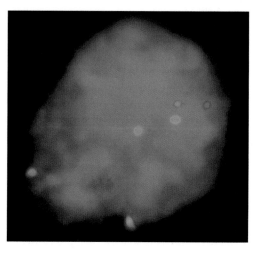

copies of chromosome 21) have long been routinely tested for during pregnancy by the technique of amniocentesis, in which foetal cells floating around in the amniotic fluid are removed and examined by microscopy. This, and a more recent complementary technique, chorionic villus sampling or CVS, are both offered routinely in western countries to mothers over the age of 35, who are more likely to produce eggs with abnormal chromosome content. A positive result enables parents to make an informed choice about whether or not to terminate the pregnancy.

More recently, however, it has been possible for couples to plan for healthy children without having to make harrowing decisions about termination. Instead they can follow a procedure known as pre-implantation genetic diagnosis, or PGD, which piggybacks on the success of 'test-tube baby' or *in vitro* fertilization (IVF) technology. About three days after fertilization with donor sperm, the egg gives rise to a six- or eight-cell stage embryo, from which one cell can be removed and analysed for the presence of abnormal chromosomes or gene mutation, without affecting the baby's development. Only normal embryos are selected and placed into the mother's womb.

PGD has now been used more than 1700 times in clinics around the world, leading to the birth of more than 300 healthy children (according to the European Society for Human Reproduction and Embryology PGD Consortium in December 2002). Embryos are screened for many different conditions, including CF, Huntington's disease, sickle-cell anaemia and Duchenne's muscular dystrophy. Checking the embryo's sex (XX for female and XY for males) may also help to avoid suspected X-linked disorders which only occur in boys because they lack a second X chromosome with which to compensate for an abnormal gene on one X chromosome.

The 'fat' epidemic

But single-gene disorders, though there are many different types, are in fact relatively rare. Far more common are the so-called 'complex' or 'polygenic' disorders, involving the combined actions of many genes. One only has to see the news headlines to know that we have a new epidemic in developed nations: obesity. This excess of body fat affects around 20 per cent of the US population, slightly fewer for Europe, and is usually attributed to modern sedentary lifestyles coupled with unhealthy fat-filled diets. Obesity in turn predisposes people to serious life-threatening complications of

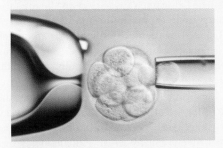

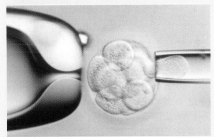

Pre-implantation Genetic Diagnosis (PGD) involves a combination of IVF and molecular biology tools. While still in the culture dish, a fertilized egg will begin dividing almost immediately, and by three days will have given rise to a six- or eight-cell embryo. At this point, geneticists can remove one of the cells by suction into a fine tube, and test for the presence of abnormal chromosomes by the technique of FISH (fluorescence *in situ* hybridization). This involves placing the cell on a microscope slide and flooding it with a tailor-made dye that makes the affected chromosome 'glow' under UV light. The dye is tagged onto DNA 'probes' – short stretches of DNA in which the nucleotide sequence is complementary to the abnormal portion of the chromosome (spanning the junction at which pieces of two different chromosomes are joined, for example).

Alternatively, the DNA can be extracted from the cell and a mutated gene detected within a matter of hours by the exquisitely sensitive PCR technique (see Chapter 3). If the embryo gets the 'all clear' it can be transferred to the mother's womb, in the knowledge that the baby will be unaffected.

Removal of one cell from a human embryo for PGD.
[courtesy of Alpesh Doshi, UCL]

type 2, or insulin-resistant, diabetes and heart disease. Perhaps most alarmingly, it is being diagnosed increasingly in children.

But obesity is not just about lifestyle, as some believe. It is to some extent the result of a far more complex interaction between the environment and our genes. This misunderstanding has particularly distressing consequences, given the social stigma attached to 'being fat'. Different people seem to be at different points along a sliding scale between being lean or fat. At one end of the scale are those who seem to be able to eat whatever they like, without getting fat. At the other end are people who, from infancy, suffer an enormous weight problem, no matter how carefully they control their calorie intake. And in the middle are those for whom there appears to be a clear link between eating a lot and being overweight.

Genetics clearly has a role: twin studies show that body weight is as much inherited as height, with genes determining up to 80 per cent of fat deposits on the body. Unlike the rare single-gene disorders described above which involve only a tiny portion of 30 000 or so genes in the human genome, obesity involves many genes and affects far more people. The same applies to the other, more common conditions, including heart disease, cancer and psychiatric disorders such as schizophrenia. They are all influenced by the environment – food, chemicals, hormones and infections – even environmental factors must at some point have their effects at the level of the gene.

These multi-gene conditions do not obey the strict Mendelian rules of 'dominant' or 'recessive' inheritance (a clear indicator of the combined effects of many genes). And besides genes, scientists are beginning to suspect that some of the variety in susceptibility between individuals may also be due to alterations in the DNA of the 'regulatory elements' or control regions that lie between genes. Using modern techniques in genetics to identify the different elements at work will lead not only to better understanding, but also possibly to new types of treatment.

Fat mice, thin kids

In December 1994, Rockefeller University scientist Jeffrey Friedman announced his discovery of a hormone made by fat tissue. Called leptin, it was missing in a mutant mouse strain that naturally develops obesity. Treating the mice with an external source of leptin reduced their appetites and restored their body weight to normal. The mice, Friedman found, carried a mutation in the leptin gene. Since then, Stephen O'Rahilly at the University of Cambridge, UK, has found that mutations in the human version of leptin account for around 5 per cent of severely obese individuals. In these, leptin treatment can bring about the most dramatic restoration of normal body weight. 'If you have the mutation the response is quite brilliant,' Friedman has commented.

Altogether, leptin treatment can provide some relief for about 15 to 20 per cent of cases of obesity 'off the street', according to Friedman. But for the remaining 80 per cent it has no effect. Nonetheless, its discovery has provided a window into a complex web of events involved in controlling body fat. One is that an increase in fat tissue leads to higher levels of leptin in the blood. These in turn trigger nerves in the brain to reduce appetite. A loss of fat tissue causes a fall in leptin and boosts appetite. Many signalling proteins are involved in these neural pathways. Friedman has said, 'Each person has an intrinsic sensitivity to leptin that sets how much fat they'll carry.'

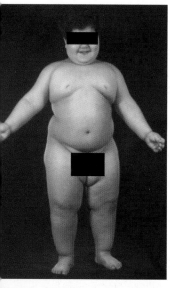

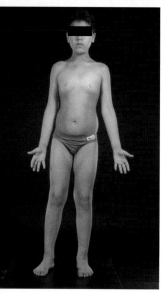

Female patient with a genetic defect predisposing to obesity before and after treatment with the hormone leptin. [courtesy of Sadaf Farooqi and Stephen O'Rahilly, Cambridge University]

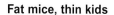

Most obese people have normal to high levels of leptin, rather than a deficiency, and are unable to respond to the leptin signal by reducing appetite. Friedman suspects mutations in other neural and hormonal pathways are responsible.

Island life

In order to probe the mystery further, Friedman is turning his attention to the Pacific island of Kosrae, in Micronesia, where there is now a high proportion of obese people, following a shift to a high fat western diet. He sees this as an opportunity to understand how our ancestors evolved so as to avoid becoming too fat or too thin:

Fat can be viewed as a bank of energy in case there's a time of famine. You really want to bank that energy in the proper amount. Too little and you risk starving to death the next time there's a famine – at least in the environment we evolved in. Too much and you risk the sequelae of obesity and, in ancient times, predation.

The inhabitants of Kosrae have a mixed ancestry, according to Friedman. The original small founder population arrived about 1000 years ago from Polynesia, and would have been at greater risk of starvation than today, encouraging a tendency towards fatness if calories became plentiful. Then they went through a 'bottleneck' in the 19th century when the native population fell from 3000 to around 300 individuals. These intermarried with around a hundred newly arrived Caucasians. Friedman hopes to use this mixing to his advantage, to tease apart the genetic contributions of the different ancestors, information that could then be applied to populations elsewhere. He has said, 'Our hypothesis might be that the genes that predispose to leanness on this island might be Caucasian in origin.'

So far, Friedman has completed a 'genome scan' of 2000 individuals, searching for microsatellites and other 'markers' (see Chapter 7) that are inherited along with obesity, in order to try then to identify the genes involved. He hopes these studies will lead to a better understanding of the complex web of events that control metabolism, and possibly new targets for therapy.

War on cancer

Since US President Nixon declared a 'war on cancer' in 1971, scientists have leapt forwards in their understanding of at least some of the many hundreds of possible mutations that lead to cancer. These can act at all control points for cell division and growth. But just when the problem seemed too bewildering to fathom, along came 'DNA microarray' technology, enabling investigators at last to look at the activities of a whole number of genes simultaneously. By placing hundreds of different gene fragments onto a glass grid, it is now possible to fish out individual messenger RNAs from an entire soup of molecules extracted from a particular tissue, by virtue of their ability to bind in a complementary way to the DNA sequence. The genes 'glow' when they bind mRNA, to produce a visual array of colourful spots, revealing which genes must have been active in the tissue to produce the mRNA that binds. In this way, investigators can see which genes are 'switched on', for example in tumour versus normal tissue, or in tumours that have spread versus those that have remained in one place.

So far, this technique can distinguish between breast cancer cells *in vitro* that spread more rapidly, and therefore would indicate which patients require more aggressive treatment. For lymphoma, a cancer of white blood cells, it also now appears possible that microarray technology could be used to predict which patients are likely to respond to different combinations of drugs. It will be some time, however, before these laboratory breakthroughs will be translated into routine clinical use.

Meanwhile, at the Wellcome Trust Sanger Institute in Cambridge, molecular biologist Mike Stratton leads the 'Cancer Genome Project', which aims to identify many new gene mutations involved in cancer. They have devised a 'high-throughput' screening method that searches the reams of sequences now available

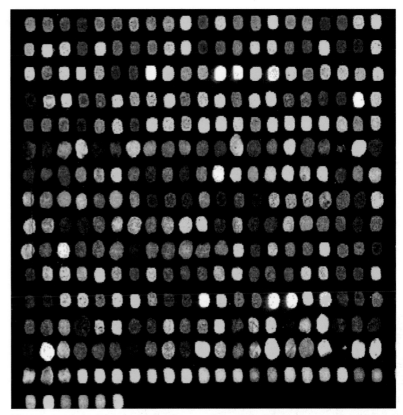

Microarray analysis: red and green fluorescent dyes reveal differences in the DNA sequence between normal and human tumour tissue, indicating the presence of deletions, mutations and amplifications associated with cancer. [courtesy of Mike Wigler, CSHL]

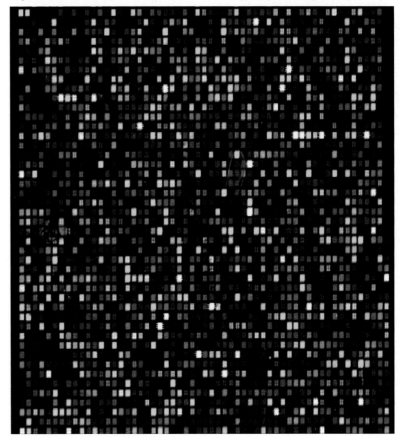

from the HGP for subtle differences in sequence between cancer cell lines and normal tissues of the same type. Within months of starting the project in 2002, the team spotted a mutation in a gene hitherto not implicated in cancer in 80 per cent of cases of people with melanoma, a skin cancer.

Smart weapons

As well as identifying gene mutations and predicting clinical outcome, scientists are hoping that advances in genetics will lead to 'smarter' drugs. For decades, we've had a sledgehammer type of approach to cancer. Cut it out, zap it with radiation and saturate it with drugs that kill all rapidly dividing cells. But these therapies are non-specific to cancer cells and all have their side effects; in many types of cancer the cells develop new mutations that enable the tumours eventually to overcome existing therapies.

The identification of new gene mutations may provide a new set of targets for future drugs that have a more tailored mode of action, with

> "The hope in cancer research is to understand the targets and come up with treatments based on that. DNA is central to every aspect of cancer."
>
> Mike Wigler, CSHL

fewer side effects. The drug Gleevec™ has already begun to point the way forwards – it is a 'small molecule' drug specifically designed to bind to and block the action of an abnormal protein that appears in a form of blood cell cancer called chronic myeloid leukaemia, or CML. For decades, this type of leukaemia has been known to contain an abnormal chromosomal rearrangement called the 'Philadelphia chromosome', in which part of chromosome 9, containing a gene called ABL, gets wrongly attached during cell division to chromosome 22 (containing the BCR gene). This causes a new gene to be made, BCR-ABL, which produces the new protein, p210. Gleevec specifically interferes with the signalling property of p210, and causes the cells to die. Clinical trials showed such an impressive effect that the US Food and Drug Administration approved its use in record time.

Bug busters

Not even microbes are safe from the gene detectives, who are now comparing the genomes of bacteria, viruses, fungi and parasites, in order to understand how they cause disease. Just as Avery, Oswald and McCarty's investigations in the 1940s (see Chapter 1) led to the crucial evidence for the role of DNA, so scientists today are comparing virulent and harmless strains of the same organism in an attempt to understand what makes the difference.

Fighting back

Scientists have already speculated about how infectious disease epidemics in the past may have provided a selection pressure for the spread of certain gene mutations in the human population. The CF mutation, for example, is suspected to have arisen in Europe centuries ago as a defence against cholera, which normally kills by causing severe fluid loss through diarrhoea. The condition is entirely absent in Africans. Yet Africans do have a high prevalence of sickle-cell anaemia, caused by a mutation in the gene for the red blood protein haemoglobin, which protects against severe malaria by causing red blood cells to change shape if invaded by parasites.

More recently, a lucky section of the Caucasian population has been discovered who appear to be resistant to HIV infection – due to a mutation in a gene called CCR5. This gene normally codes for one of the protein receptors that HIV uses to gain entry into white blood cells. A person with two copies of the mutant gene has no CCR5 protein. The mutation is thought to have arisen in the human population about 2000 years ago, and the incidence of people carrying a copy of the gene to have risen sharply from 1 in 40 000 to 1 in 5 people about 700 years ago. The date coincides, some suggest, with the timing of the Black Death, the bubonic plague that swept through Europe between 1347 and 1351, killing more than 25 million people.

Comparing the genomes of different strains of pathogens now may help scientists to understand further how they cause disease, and to predict how these organisms may evolve in future, and how we might combat the diseases they bring.

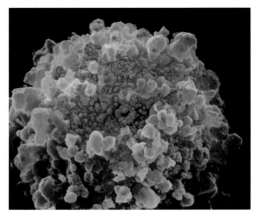

Right: Coloured scanning electron micrograph of a T lymphocyte blood cell (green) infected with human immunodeficiency virus (HIV) (red) budding away from the cell surface. [NIBSC / Science Photo Library]

DNA in culture

No longer the arcane and privileged scene that it was in the 1950s, science is now reaching every section of society, and speaks to people of all ages. No one can ignore the headlines announcing the latest news about new drugs for cancer or vaccines for infectious diseases, foretelling a world in which 'designer babies' and human clones are the norm, warning of the perils of 'mad cow disease' and AIDS, or talking of genetically modified organisms.

Underlying all these stories in the media is, at some level, research involving DNA: investigating new clues about a disease, producing a vaccine or drug in pure form, and providing the tools for creating genetically altered life forms – whether agricultural or animal. The impact of DNA and genetics is being felt everywhere.

Green fluorescent protein-tagged DNA in mice, by Eduardo Kac, from *The Eighth Day*, 2001.
Eduardo Kac, *The Eighth Day*, 2001 (detail). Transgenic installation with biological robot (biobot), GFP plants, GFP amoebae, GFP fish, GFP mice, audio, video, Internet. [courtesy Institute for Studies in the Arts, Arizona State University, Tempe]

Besides being an ever-growing source of stories for the media, research involving DNA is now enmeshed in the often bizarre story lines of many books and films, as *Nature* biology editor Christopher Surridge describes.

DNA in popular fiction by Christopher Surridge

In the 50 years since the first unveiling of DNA's structure, the image of a double helix has become ingrained in the public consciousness as a shorthand for modern biology, genetic engineering and evolution itself. Its intertwining has been used to advertise technological advances in such non-biological products as cars and digital cameras. In 1995 Bijan even brought out a perfume called DNA, although there was no DNA in it. In fiction, the notion that each cell of an organism contains a unique script driving its development from a single cell to a fully formed individual has provided much scope for novel plots or at least new twists to old ones.

Of all the advances in molecular biology, cloning – the creation of genetically identical individuals – is the most visible in fiction. Thus, many years before a Finn Dorset lamb called

Dolly showed this to be possible, dinosaurs were returned to the Earth in Michael Crichton's 1980 novel *Jurassic Park*, which inspired the Steven Spielberg epic of the same name. The dino-DNA came from mosquitoes, preserved in amber just after feeding, but was fragmentary and so had to be stitched together with the genomes of modern reptiles. The motive for this audacious project: stocking an exotic theme park.

Of course, if animals can be cloned then so can humans. The fictional justifications for this are many and varied. In the film comedy *Multiplicity* (1996), Doug Kinney (played by Michael Keaton) clones himself as a labour-saving tactic, enabling him to be in two, three and even four places at once. More sinister reasons lie behind Arnold Schwartzenegger's duplication in *The 6th Day* (2000), where laws banning human cloning, while allowing it in

Jurassic Park (US 1993).

[Universal/Amblin/Ronald Grant]

family pets, are systematically broken so that the clones can take over the world.

Cloning has also provided authors with a way of bypassing death. The series of *Alien* films used it to 'resurrect' its leading human character, Ellen Ripley (Sigourney Weaver), killed at the end of the third film and ready for further battle in the fourth – aptly named *Alien: Resurrection* (1997). But one of the earliest, and still best-known, uses of human cloning in film was to duplicate Adolf Hitler in *The Boys from Brazil* (1978). This was one of the few examples of the 'cloning' genre to address carefully the question of how much a person's personality is shaped by DNA rather than by experience. The ageing Nazis behind the scheme covertly attempted to make the developing clones' environments as similar as possible to that of the original Hitler, including the murder of their 'fathers' at the age of 65. It was these assassinations that ultimately led to the plot being discovered.

Totalitarian overtones also colour another use for cloning in fiction. The coupling of cloning with genetic engineering can apparently create armies of identical soldiers, specially adapted for combat. The character of the Rogue Trooper in the UK comic *2000 AD* is paradoxically left alone and isolated after the slaughter of all his clone-brothers. On film this idea has been used less thoughtfully in *Universal Soldier* (1992) and recently in *Star Wars Episode II: The Attack of the Clones* (2002).

Despite their nods to modern scientific discoveries, these examples generally tell a very old story about the dangers of 'tampering with nature'. This dates back to the dawn of science and Icarus flying too close to the sun, melting the wax that secured his wings and falling to his death. Aldous Huxley's *Brave New World* (1932) considers the consequences of social and genetic engineering while H. G. Wells's *The Island of Doctor Moreau* (1896) and Mary Shelley's *Frankenstein* (1818) explore the moral and ethical pitfalls of creating life. All three were written well before 1953.

A more prosaic, though no less fundamental, use of DNA occurs in the film *Gattaca* (1997). Here the sequencing of a person's DNA allows their genetic propensities for disease, physical and mental abilities – even personality – to be rapidly determined from as little as a pinprick of blood. The result is a caste-ridden society based on genetic screening – a free market capitalist version of Huxley's Brave New World. But against this background the hero, Vincent Freeman (Ethan Hawke), proves that inheritance isn't everything, cheating the genetic tests

and achieving what his DNA said was highly improbable.

In so-called 'hard' science fiction Greg Bear most tightly weaves molecular biology into his novels. In *Blood Music* (1985) a molecular biologist uses the redundant or junk DNA between genes to build tiny computers which, true to the spirit of Icarus, go on to destroy the world as we know it. In *Darwin's Radio* (1999) a pandemic, Herod's influenza, is causing hideous birth defects through activation of retroviruses that have lain dormant within the human genome for millions of years. It transpires that this is not a purely hostile plague but rather an instrument of evolution shuffling the human genome to allow mankind to become a new species altogether.

Darwin's Radio is also unusual as its central character is a working and recognizable molecular biologist. This remains something rare in fiction. In campus novels such as those of Malcolm Bradbury, David Lodge and Kingsley Amis, the academics are almost exclusively drawn from the humanities – with the notable exceptions of David Lodge's cognitive neuroscientist in *Thinks* (2001) and Angus Wilson's earlier plant biologist hero in *As if by Magic* (1973). Carl Djerassi has made some attempts to place the modern molecular biologist centre stage in his self-styled 'fictional science' novels and plays; however the most likely genre in which to find a character who understands the possibilities of DNA is detective fiction.

The rise of forensic scientists in the crime novel literature has been slow, but they are now established members of the supporting cast. In the novels of Patricia Cornwell, the heroine, Dr Kay Scarpetta, is a pathologist. Even so, it is hard to find any occurrence where DNA fingerprinting, in which the unique patterns within an individual's genome identify them, is used to any great effect. Perhaps the lack of uncertainty that such a technique provides, if it can be applied, is anathema to the dramatic tension of the whodunnit.

There is, however, a use of DNA for identification that appears time and time again: the paternity test. While previous generations of authors have required familial resemblances, rare blood groups, birthmarks or the testimonies of ancient nursemaids to sort out the tangled parentage of their characters, these days a simple blood test will resolve the matter within weeks. Given that the plot line of a child with unknown or disputed parentage is a staple of ongoing television and radio 'soaps', reference to a DNA-based paternity test are broadcast to untold millions of households almost weekly. Here at least fiction has not overlooked DNA.

One of the strongest and most enduring symbols of the impact of new developments in society occurs when they become a source of either inspiration, or new material in the art world, as science journalist Steve Nadis reveals.

DNA as an art form by Steve Nadis

The Human Genome Project is, in the most general sense, a search for identity – a quest to discover the features that make *Homo sapiens* unique. A broad enquiry like this, which is trying to reveal 'the Book of Life', as some have put it, is clearly not the exclusive domain of molecular biologists. It's not surprising, therefore, that artists have jumped into the fray, exploring the realm of genes and DNA from a wholly different perspective than that of their scientific contemporaries. Some of the artists in this field have used genetics as the subject of their work with micrographs of chromosomes and artistic renderings of the genetic code, while others have incorporated DNA directly into their creations. Regardless of how these endeavours are ultimately judged, they have already stretched the boundaries – and raised new possibilities – as to what constitutes art.

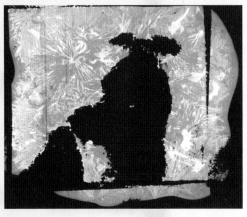

Fluorovenus by Joe Davis.
[Joe Davis]

Joe Davis, artist-in-residence in the Massachusetts Institute of Technology's biology department, is a pioneer in this field. Working with Harvard Medical School biologist Dana Boyd, Davis synthesized his first DNA 'sculpture', called *Microvenus*, in 1987, inserting a customized DNA sequence into the genes of *E. coli* bacteria, which obliged by reproducing his design a trillion-fold. When properly deciphered, the sequence provides a symbolic rendering of the female genitalia – a reaction to the sanitized depictions of humans carried onboard the Pioneer and Voyager space probes. Microvenus-laden bacteria were first presented to the public in petri dishes, locked behind a glass refrigerator door, at the 2000 Ars Electronica festival in Linz, Austria. In a nearby exhibit, Davis displayed another DNA-encoded message in a petri dish, which was labelled: 'I am the riddle of life. Know me and you will know yourself.' He is partway through a more ambitious, 3867-base-pair DNA synthesis (among the largest syntheses ever attempted) that would re-create, in a schematic way, an infrared image of our galaxy captured by the Cosmic Background Explorer (COBE) satellite.

Genesis, another project in this vein, was conceived by Eduardo Kac and unveiled in 1999. A Brazilian-born artist based at the Art Institute of Chicago, Kac translated a passage from the book of Genesis into Morse code, which he then converted into a DNA sequence that was implanted in a bacterial gene. A more recent effort, *The Eighth Day*, was unveiled in 2001. It brings together four newly created transgenic animals – a mouse, fish, plant and amoeba – housed within a plexiglass dome. Each animal contains genes that code for green fluorescent protein. 'It's a complete fluorescent ecology,' Kac says, 'different beings that didn't exist before co-existing together.' The title of the piece suggests that creation did not end with the seven days described in the Bible. Indeed, Kac promises that his future artworks will involve the invention of new life forms.

Davis is also embarking on his first genetic manipulation of a higher organism, the fruitfly, by writing a sentence from the Greek philosopher Heraclitus into a fruitfly gene that regulates eye colour. But he hopes to make this 'the most environmentally friendly manipulation

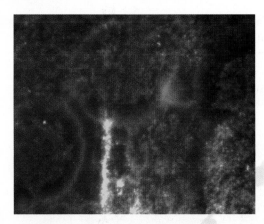

ever done' by employing the so-called 'silent code' he developed with Boyd. The technique 'exploits the degeneracy of the genetic code' – the fact that there are multiple ways to make each amino acid – 'to put information into existing genes without changing their biological function,' Boyd explains. 'Basically, we're inserting information through a loophole in nature.'

Despite his current preoccupation with the silent code, Davis still creates more conventional works such as pictures and paintings. Recently, for instance, he devised a technique called DNAgraphy that uses the DNA of streptococcus protein as a key ingredient of the emulsion for developing photographic film. The theoretical resolution with this approach is much higher than with standard silver halide solutions, he claims, making more detailed pictures possible. As part of his 'living colour' project, Davis prepared yellow, green, red and cyan 'paints' by inserting fluorescent proteins into *E. coli* plasmids. In 2001 Al Wunderlich, an abstract artist who teaches at the Rhode Island School of Design, used these pigments and a fine-edged brush to make four tiny paintings, each roughly the size of a quarter.

Wunderlich enjoys the spontaneity of putting live paint down, watching it under the microscope, and then responding to the changes. The project reminds him of the early, free-form days of the abstract expressionist movement in 1950s and 1960s New York, 'when the whole point was that painting was supposed to be a living experience'.

David Kremers, a conceptual artist at the California Institute of Technology, has employed a modified version of this approach for about a decade. Instead of using a brush, he grows bacteria genetically engineered to make coloured proteins – in layered patterns on acrylic plates. 'I essentially collaborate with the bacteria, which take about 16 hours to grow in ways that I encourage,' he says. 'Then I put them in suspended animation to preserve the image.'

Kremers is critical of much of the efforts that currently pass for 'genetic art'. Many of the pieces are not fully realized artworks because they are dominated by technology, he says. 'They may be technically seductive without asking the deep questions that art is really about.' This problem will persist, he adds, until artists become more proficient in the tools of molecular biology. 'You really can't focus on the art until you've mastered the technology.'

Nevertheless, Davis believes this new movement is 'the future of art', not a fad, because profound advances in biology will affect every aspect of life – including, especially, art.

Kac agrees that genetic or transgenic art is 'here to stay', adding 'It's inevitable that this mode of working will be incorporated into the larger vocabulary of art.' Ironically, he says, 'as transgenic art becomes more established, it will become less visible – just another tool that artists can draw on'.

Although the field is definitely growing, it's too early to tell whether the current wave of activity will yield enduring works of art. 'You can prove all sorts of things in science, but you can't prove anything in art,' Kremers notes. 'Time is the only real measure. One hundred years from now, people will only be talking about the important works.' It remains to be seen whether 22nd-century critics will be talking about artwork that incorporates DNA, modified genes or genetically engineered organisms.

Left: *Emerge* by Al Wunderlich. [Al Wunderlich]

Somites by David Kremers, gesso, agar, x-gal, iptg, anpicillan, coomassie blue r-250, ecoli tb-1, plasmid, synthetic resin on acrylic plate, 24 x 24 inches, 1992. [collection Carolyn Kremers]

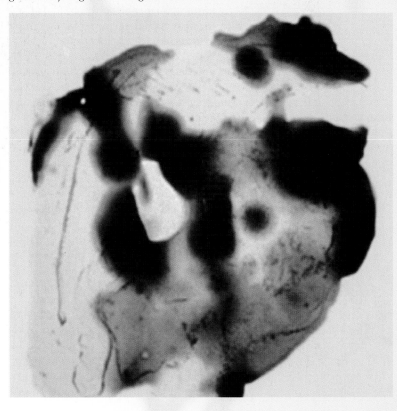

"The kids are very open-minded. At the end of these kinds of experiments they ask us questions. They're not aware of the scientific work in the literature, and so they're usually asking the key questions that a lot of the scientists are asking themselves."

Scott Bronson

[Julie Clayton]

School students at the Dolan DNA Learning Center, CSHL. [Dolan DNA Learning Center, CSHL]

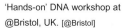
'Hands-on' DNA workshop at @Bristol, UK. [@Bristol]

DNA in education

On Long Island, New York, more than 22 000 school children a year troop through the doors of the Dolan DNA Learning Center, past a replica of Watson and Crick's brass plates model of DNA in the foyer, into a fully equipped laboratory where they can chop up DNA using the same kind of 'restriction enzymes' that helped to create the biotechnology revolution of the 1970s, or into an interactive computer suite where they can pursue their own investigation of human ancestry.

They are just ordinary children, with all manner of aspirations. Only a minority are likely to pursue science to university level or beyond, but they are all going to be affected by genetics at some stage in their lives. An awareness when young of the basic elements that underlie the subject is vital, if they are to progress in a world where they will need to make informed decisions about their health and their food, and take part in public debates on where the research should go next.

Scott Bronson, a tutor at the Learning Center since 1996, has commented 'If 5 per cent are interested in pursuing a career in the sciences I would be surprised. But the field of biology has really become quite broad now … It's important to make them slightly more literate in science … If we want individuals to be responsible, they're going to need to be informed citizens – it's to their advantage.' Scientific knowledge presented in this form can also broaden children's awareness of potential career options. 'There are a lot of kids that don't like science too. It gives them a new way of looking at science. A kid who's interested in computers and mathematics now has a huge role to play in biology, as the genome projects need biomathematicians.'

Rather than teaching a list of facts, the emphasis of the lab experience is to convey a sense of the process of science, to stimulate what the centre's director David Micklos calls 'a real debate about science' in the kind of discussions that can take place in the home. 'Parents are thrilled … The kids go home and say, "Hey, I put DNA into bacteria today in the lab," and the parents say "You did?" and the kids say, "Yeah, it was real easy, it just took 45 minutes." That's what generates a true discussion about these things … It's probably the best way that you can actually get some adults to be semi-literate about these things.'

These workshops are a clear illustration of how experiments that were once carried out within the confines of high tech laboratories are now widely accessible. In the UK, school children from all over the south and southwest of England flock to At-Bristol, a hands-on science activity centre in Bristol, where they can manipulate their own DNA, taken from a scraping of their own cheek cells. They copy the DNA, using the polymerase chain reaction (PCR) technique invented by Kary Mullis (see Chapter 3), and then participate in debates about the uses of DNA in society – for example in genetic testing, cloning and stem cell research.

The genie is out

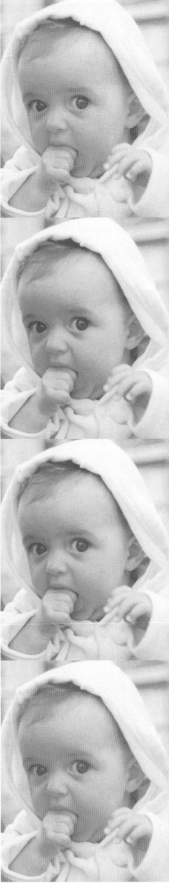

Year of the clone?

The year 2003 began with an extraordinary but unconfirmed claim of the creation of a human clone by Clonaid, a US company that offers gullible customers eternal life through cloning, on behalf of a religious sect called the Raelian Movement. Meanwhile, Italian fertility specialist Severino Antinori had also outraged both scientists and the media by his claims during 2002 that several clonal pregnancies were under way, again without independent verification. Even if the claims turn out to be false, there are bound to be similar ones in future. To some people the claims are mere publicity stunts, while to others they are an abuse of human rights. Either way, they illustrate the impact that genetics is now having on society.

As with the advent of many new technologies, the first reactions to the idea of human cloning are a mixture of fear and moral outrage. Scientists and ethicists have wasted no time in calling for debate about the potential harm to society of cloning technology, so that safeguards will be in place should the possibility become real. Such fears over the power of genetics are not new. Ever since the 1973 Asilomar meeting led to a two-year ban on recombinant DNA technology in the United States, scientists have found themselves accused of 'playing God' as they attempt to manipulate the DNA of living organisms. And in recogniz-ing the need for social responsibility, James Watson, for example, insisted that the HGP also include research into social, legal and ethical implications raised by genome sequencing.

Beyond reality

The hype surrounding human cloning increased exponentially following the announcement in 1996 of the cloning of Dolly the sheep by scientists at the Roslin Institute in Scotland. Dolly was a clone of another adult sheep, in which the nucleus of an adult cell, containing a complete copy of the animal's genes, was inserted into an egg which had had its nucleus removed, and the egg was jolted into cell division by applying an electrical current. The resulting embryo was then implanted into the womb of a ewe. Although it was the culmination of a long line of incremental advances in the agricultural science of farm animals, the creation of Dolly caused an overnight sensation on account of its implications for human cloning.

Indeed, Lee Silver, a molecular biologist and author at Princeton University described recently how the public perception of cloning has leapt far beyond any sense of reality:

I was recently contacted by a Dutch television pro-ducer for my reaction to reports that a fringe reli-gious group was ready to use cloning to bring dead

"Just as most people still want to believe that there is something more extraor-dinary than molecules at the heart of living existence, they cannot easily accept that DNA is all that important, being the crucial difference between the various forms of life, without endowing it with virtual genie-like properties that we will never be totally able to handle."

James Watson, *Passion for DNA*

[Julie Clayton]

Dolly, who died in February 2003. [Roslin Institute, Edinburgh]

children back to life. For the umpteenth time, I explained that no technology exists for making copies of people, and that real cloning technology might only lead to the birth of a unique and unpredictable child who had the same DNA sequence as someone else, but nothing more. The producer was abrupt and dismissive: 'Dr Silver, you are not aware of what cloning can accomplish. Clones are not what you think they are.'

(*Nature*, 412: 21, 2001)

Scientists – including Ian Wilmut, Dolly's creator – have frequently warned against technical obstacles to human cloning. They cite numerous failed attempts at cloning animals,

> "It is concern over these issues that makes me and many others reject the suggestion of cloning a person"
>
> Ian Wilmut, creator of Dolly the sheep

including cows and mice as well as sheep, and the physical defects such as heart and lung problems and deformities. Dolly, for example, developed arthritis prematurely, and died in February 2003 at the age of six years, only half the normal lifespan of a sheep.

Many scientists accept such limitations for farmed animals, as John Sulston, one of the leaders of the HGP, has explained: 'there's nothing wrong with this because it's not actually that it's particularly inhumane. They may have short lives but that doesn't mean they're un-happy ones. But they can provide breeding stock to produce new animals that are carrying new genes.' In contrast, 'technically to produce a human being by cloning is just like murder or child abuse – you just don't allow it'.

These are Sulston's personal views, but he is also a member of the UK's Human Genetics Commission, which debates these and other issues pertaining to the potential uses and abuses of genetics technology. These debates are important, believes Sulston, because human cloning may one day be possible, and society does have to 'consider all these things rationally' rather than to make decisions based on fear.

Identity crisis

If, in future, human cloning becomes technically safe and feasible, what would the social consequences be? After all, genetically speaking, it would be no different to having an identical twin – already one of 'nature's clones'. Environmental effects and life experience would ensure that the two or more cloned individuals would have (at least partly) different personalities. But what about their sense of personal identity? How might a mother relate to her 'son' who is actually his father's clone, but twenty years younger, fitter and more attractive? How might a daughter be affected psychologically as an adult, knowing she is her mother's clone, particularly if she was brought into the world to replace a dead sister? What would be her parents' and teacher's expectations? And what would be the pressures on a child who knows that the parent from whom they are cloned comes to suffer from a disease for which the child's own organs are the only hope of a long-term treatment or cure?

'It is concern over these issues that makes me and many others reject the suggestion of cloning a person', wrote Ian Wilmut, creator of Dolly the sheep (*Nature*, 412: 583, 2001).

Cloning permits

In the UK, the furore over the possibility of human cloning has heightened awareness of the need to pass new legislation to control the technology and to pass ethical regulations about research and clinical practice. As a result, in November 2001 the UK parliament passed a Bill that allows 'therapeutic cloning', while at the same time banning 'reproductive cloning'. This means that scientists can apply for a licence to create cloned human embryos in a culture dish for the purpose of investigating new avenues for treating human disease – for example the creation of human stem cells for the repair or replacement of damaged organs – but are not permitted to implant the embryo into a woman. In the US, however, and in most of Europe, therapeutic cloning is not permitted either; scientists are obliged to use a limited selection of pre-existing stem cell lines, many of which have been criticized for being of questionable quality.

Designer babies

A more immediate possibility than human cloning, however, but getting less attention, is the use of recombinant DNA technology to tinker with the genetic make-up of normal babies. This may either be for preventing future disease, or to boost desirable attributes. It only needs a gene therapy-like insertion of a foreign gene into a human egg plus IVF, and a pre-implantation screen to ensure that the new gene is in place, and hey presto – a child with permanently altered DNA. And because the new gene(s) would also exist in the 'germline' – the sperm and egg-forming tissues of the ovary and the testes – they would pass on to all future generations.

As for cloning, questions are being raised about the ethics of 'genetic enhancement'. Should it ever be allowed, and for what pur-pose? But the more we understand about our genes and the roles they play in disease, the more we may be tempted to tamper with them in future, despite the risk of side effects. As James Watson asks in the accompanying interview, who wouldn't want to protect their children from developing cancer in later life? Perhaps societies need instead to consider drawing a line between allowing genetic enhancement for the purpose of preventing disease, and modifications 'on a whim' that affect other traits such as personality and aptitude.

Safeguards

In the UK, no such 'designer baby' procedure is permitted, and individually gene therapy, IVF and PGD are strictly regulated. But that does not necessarily stop gene tinkerers elsewhere from having a go, even if the success rate of IVF remains at only 20 to 30 per cent, particularly as plenty of potential parents may demand it.

Fear of discrimination

Advances in genetics have also led to concerns of a different kind, that genetic testing for new disease-related mutations could lead to discrimination by employers and health insurers. This concern has prevailed since the 1970s when US authorities began a campaign of screening for sickle-cell anaemia in African Americans. People who were found to be carriers of a

[Julie Clayton]

sickle-cell mutation were mistakenly assumed to have the disease, and found themselves unable to get health insurance. In some cases they were refused employment.

The Association of British Insurers has responded to such fears with a 5-year moratorium on the use of genetic test information to assess insurance premiums, with the exception of Huntington's disease, and in the US many individual states have enacted their own laws to prevent discrimination. But insurance companies in the US are opposing such restrictions on the grounds that they create an 'information asymmetry' which prevents accurate assessment of risk and spreading the cost of insurance premiums accordingly over the entire population. They also argue that the worry about discrimination is groundless, with little or no evidence that it occurs today. Nonetheless, the fear itself has apparently been enough to stop some people from seeking genetic tests, even if they have a family history of a condition. This is especially the case for conditions which have no proven treatment, such as cancer, where a test offers little advantage to the individual.

Dealing with risk

Even without discrimination, many people would prefer to live in ignorance rather than face a test result that gives no certainty about developing a particular disease. An example here would be the tests for breast cancer-related mutations in the genes known as BRCA1 and

"Most of these genes that predispose to one disease or another are not going to invariably predict the outcome, they'll just change likelihoods."

Jeff Friedman, Rockefeller University

BRCA2. These account for just a small minority of breast cancer cases, so a negative result cannot give total reassurance, while a positive result does not mean that a woman will definitely get the cancer. In future, genetic testing is increasingly likely to include genes of this kind, which play a small role in a complex disease, including in other conditions such as obesity and heart disease. Rockefeller University's Jeff Friedman has explained, 'Most of these genes that predispose to one disease or another are not going to invariably predict the outcome, they'll just change likelihoods.'

But might some people pursue genetic tests anyway, even with an uncertain outcome? Friedman believes that people may choose to be tested anyway, because at least they could

attempt to minimize any potential harm by changing their lifestyles. If this notion catches on, there could be a surge in the number of people seeking genetic testing. Some companies are already anticipating this trend and offering gene testing via mail order or even, as attempted in the UK recently, over-the-counter at high street shops.

Should this happen, people could get results without ever having visited their physician for a professional interpretation of the results. The HGP serves only to enhance this possibility, but a physician's interpretation, and appropriate counselling, is essential, according to Friedman:

The genome project has the potential to provide hundreds of thousands of such tests with complex information that's going to elude the grasp of any one individual's ability to retain it. So some mechanisms are going to have to be developed to allow this information in time to be officially and properly interpreted.

But, as Friedman points out, there are not enough genetic counsellors to do the job properly.

Control freaks

In a world full of discoveries and inventions, it is easy to think that we can gain the upper hand over adversity. For millennia we've had fire and clothes to keep us warm, and now central heating and air conditioning to help us survive conditions too hostile for our ancestors. We can even go to the moon and back without getting frostbite. So why not assume that we can also take control of our genes?

In 2003 we are celebrating not only the discovery of the double helix, but also the creation of new tools with which to improve our lives. Until now, we have used medical technology to defy death and disease, and to enable women to conceive babies well beyond the time when fertility begins naturally to decline, so using genetic engineering to our advantage is simply the next logical step. Successes are already evident with gene therapy for children with defective immune systems.

But because of the inherent dangers of the technology, society needs to protect itself by making rational informed decisions about how to proceed. At the same time, now is the best time to put the fear of the genie back in the bottle.

In Part II of an exclusive interview, James Watson looks forwards to the next 50 years when scientists are likely to gain a far deeper insight into what controls human behaviour, and controversially, argues in favour of 'designer babies'.

INTERVIEW WITH JAMES WATSON

Part II of James Watson interview December 2002

To what extent is human behaviour genetically driven?

We're really asking what is an animal, or in this case, what is a vertebrate? Probably our most dominant motion is to eat, and after that it's sex. So we shouldn't be surprised to hear men think about women about 90 per cent of the time – I can't speak for women.

To what extent do you think this is genetically driven?

I think it's purely genetic. We're driven to it. And if you're really hungry you stop thinking about sex because sex doesn't do you any good if you're dead.

What about violence and anti-social behaviour?

It's sometimes necessary for survival – if someone's going to hurt you, you have to respond. Many animals are violent. You can't say it's wrong if it's part of animal existence.

But isn't there a danger that in future a genetic definition would become an excuse for anti-social behaviour?

I think we would understand it better but I don't think we would tolerate it differently. If someone were born a killer you would probably have to lock him up … I think that over this century we will come to grips with the nature of individual differences in what we do. How do you handle those differences? You can say, some of these people have characters I don't like. I think we'll understand a bit better but it doesn't take finding new genes to know that someone's a shit. The genes won't cause discrimination – discrimination comes because you don't like them, you don't want that sort of person in the room.

People can, of course, change and adapt their behaviour …

Of course. On the other hand, most girls who are born wallflowers, or the equivalent, are probably going to stay that way for life – they don't have much choice.

You don't think that's to do with upbringing?

Psychologists when they study twins say it has very little to do with upbringing. You read that personality is not influenced much by upbringing.

Will we become over-dependent on genetics for a quick fix for medical problems where lifestyle change might be more effective?

With lifestyle you can control diabetes, but sometimes you need insulin. I think we're just going to find out what genes do.

But do we need high-tech medicine to deal with problems like obesity?

I just want people to be able to enjoy food and not get fat. Some people can eat almost anything and not get fat while other people … it's in their genes. People are terribly prejudiced against obesity but it's very hard to control and it would be nice if we could understand the science so we could develop a food without fat – like sex without babies, the revolution of the 1960s. Some people would say that was immoral – you shouldn't eat food and then just take a pill to get rid of it when they are starving somewhere, but on the other hand what's wrong with having three scoops of chocolate ice cream?

I suppose a different kind of example would be fitness – that some people say it would do you far more good for avoiding cancer and heart disease if you had a half hour walk every day …

We don't know how to avoid cancer – it's not that under control. And some people are just

born with more artery disease. They're going to get high blood pressure and they're going to get strokes, and other people who live exactly the same way won't.

You can't just feel anxious all the time about it [lifestyle]. At this stage in my life I'm an exercise freak, so I'm not saying this is bad but I'm saying that … a lot of people get bad throws of the genetic dice. They have arthritis, or they feel sickly all their lives. I don't think it's psychological, I think it's bad throws. I see our role as trying to bring the message of genetics that there's a vast amount of inequality and I think we should try and redress it, instead of complaining that we know there's inequality and that it's all our fault in how we live our lives – two different viewpoints.

How can we address that inequality?

You can genetically enhance people; if you give a new gene so that you don't get arthritis, is that bad? You could say that we don't have to do that – people can just take a pill – maybe in some cases you can't do it, I don't know. I always win the argument about using germline therapy by saying 'Everyone knows the Irish people are stupid!' Could you make the Irish better? How much is in the genes? And how much is it [a question of whether] 'if you only got rid of the Catholic faith would the Irish be better off?'

Do you have Irish ancestors?

Yes. My grandmother is a Gleeson, so on one side I have an Irish cultural heritage. I like the Irish very much.

I don't think there's anything wrong with it [genetic enhancement]. We say you shouldn't have designer babies – well, women want designer clothes. It's more or less saying you should accept what you get. For a long time people were against plastic surgery but now most people want plastic surgery.

Do you think the day will come when genetic enhancement will become possible?

It may always be impossible, but to the extent that intelligence is inherited, if you could make your children more intelligent wouldn't you do it? What's wrong with it? You could say, well, you want to make them kinder … And if you could design your children so that they didn't have cancer would that be any different?

But surely protecting from disease so that they can have a healthy life is quite different to wanting to change their personalities …

No, I don't think so. Who wants a rude daughter? That's not very nice for the daughter.

But then wouldn't we all end up being too similar in our abilities?

No, we'll always be unique.

Wouldn't it exacerbate inequality for those who cannot afford to pay?

No. Again I think this is the assumption that we spend all our [research] money making the healthy healthier rather than making the unhealthy healthy. It's a view on society, where basically I think that society is more good than bad, and that we really do care about the unfortunate … The wealthy were the first people to have television. It's again a question of cost. We're not trying to produce designer babies, we're trying to produce people who are less unequal.

Are there any areas that concern you where genetics could be misapplied?

Yes – trying to prove that Prince Harry is illegitimate. That's wicked.

What about use of national DNA databases for crime prevention?

I'm all in favour. People say it would be a police state, but now we have hidden cameras in every shopping centre – is that good or bad? If it catches someone who's killed someone, it's good. Should you be able to abandon your wife and four kids and go off and change your identity? All of us have seen this Central Park jogger case in New York where the justice was pretty wretched, and it was only DNA that showed that they weren't the rapists – these four black kids. All of them had been convicted and they weren't guilty at all. Criminal justice is still very unfair at times, and DNA will make it safer.

When you were celebrating the discovery of the double helix did you have a sense then that you might be 'letting the genie out of the bottle' and into the wrong hands?

No I've never felt that. Biological warfare could be made worse – you could say that we could make worse smallpox, but smallpox without that would be nasty. Science can always be used for harm. The question is, have we made our lives better over the past hundred years? I would say yes, and I would think that over the next hundred years we're going to make it better still. I think a crucial disaster is likely to be an infectious disease that might come back. Imagine if something killed half the people in the world: we'd be in a recession for years. I believe that knowledge is a good thing and that people will, for the most part, try and use knowledge constructively, but there will be future Hitlers and Stalins, and Idi Amins, and so on.

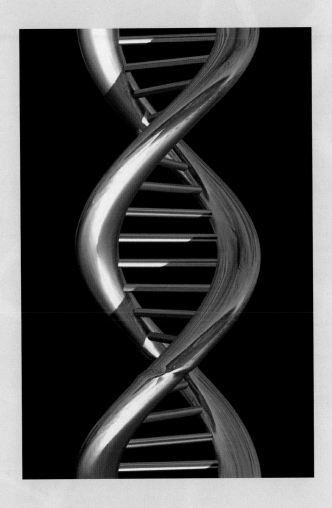

The research papers

The following section is reprinted from the pages of the special issue of *Nature* published as a supplement on 23 January 2003. That supplement marked the 50th anniversary of the discovery of the DNA double helix.

The section begins with an introduction to set the context, by Carina Dennis and Philip Campbell. It is followed by facsimiles of the three original papers published in *Nature* on 25 April 1953: by James Watson and Francis Crick, Maurice Wilkins and his co-authors, and Rosalind Franklin and Ray Gosling. The next three essays present the historical perspective, and are followed by four papers that explore the significance of DNA in medicine and in culture. The final group of papers focus on progress in biological research since the discovery of DNA, bringing the revelations of 1953 into the present and also into the future.

The eternal molecule

As a prelude to the many celebrations around the world saluting the 50th anniversary of the discovery of the DNA double helix, *Nature* presents a collection of overviews that celebrate the historical, scientific and cultural impacts of a revelatory molecular structure.

Few molecules captivate like DNA. It enthrals scientists, inspires artists, and challenges society. It is, in every sense, a modern icon. A defining moment for DNA research was the discovery of its structure half a century ago. On 25 April 1953, in an article in *Nature*, James Watson and Francis Crick described the entwined embrace of two strands of deoxyribonucleic acid. In doing so, they provided the foundation for understanding molecular damage and repair, replication and inheritance of genetic material, and the diversity and evolution of species.

The broad influence of the double helix is reflected in this collection of articles. Experts from a diverse range of disciplines discuss the impact of the discovery on biology, culture, and applications ranging from medicine to nanotechnology. To help the reader fully appreciate how far the double helix has travelled, we also include the original landmark paper by Watson and Crick and the two accompanying papers by Maurice Wilkins, who shared the Nobel Prize with Watson and Crick in 1962, and by co-discoverer Rosalind Franklin, and their co-authors (pages 83–87).

Transforming science

Given the immense significance of the double helix, it is difficult to imagine a world that wasn't transfixed by its discovery. Yet, as Robert Olby recalls on page 88, the proposed structure initially received a lukewarm reception. Maclyn McCarty, who, together with Oswald Avery and Colin MacLeod, had previously showed DNA to be the substance of inheritance, shares his personal perspective (page 92).

In science, where a lifetime's work can often be encapsulated in a few shining moments, the greatest controversies are sometimes over the sharing of credit. The discovery of the double helix is no exception. The premature death and posthumous treatment of Rosalind Franklin, whose X-ray images of DNA fibres revealed telltale clues of a double helical structure, propelled her portrayal as a feminist icon. But, as discussed here by her biographer Brenda Maddox (page 93), Franklin is better remembered as a committed and exacting scientist who saw no boundaries between everyday life and science.

Most of our readers will have grown up with the double helix, and yet it is still startling to consider how quickly DNA biology has progressed in just a lifetime. Bruce Alberts reviews how the elegant pairing of the two strands of the double helix revealed the mechanism for replicating the essential units of inheritance (page 117). Errol Friedberg considers the vulnerability of the DNA molecule to damage and the multitude of ways in which cells repair the damage (page 122). And Gary Felsenfeld and Mark Groudine describe how the gargantuan DNA molecule is packaged inside the minuscule cells of the body, and how an additional layer of information is encrypted within the proteins intimately associated with DNA (page 134). It is perhaps salutary also to recognize what is still to be learnt about the physiological states in which DNA exists, as discussed by Philip Ball (page 107).

As reviewed by Leroy Hood and David Galas (page 130), DNA science generated the tools that spawned the biotechnology revolution. It enabled the cloning of individual genes, the sequencing of whole genomes and, with the application of computer science, transformed the nature and interactions of molecules into an information science. Carlos Bustamante and co-authors consider how we are still learning much about the distinct structural and physical properties of the molecule (page 109). And according to Nadrian Seeman, DNA may develop new applications as a material for nanoscale engineering (page 113).

Influencing society

Beyond scientific and technological forums, the double helix has imprinted on society's views of history, medicine and art. As discussed by Svante Pääbo (page 95), the records of evolution have been recalibrated with information traced through DNA sequence. On page 98, Aravinda Chakravarti and Peter Little revisit the 'nature versus nurture' debate and our developing view of the interplay between genetic and environmental factors in human disease. And DNA science will transform clinical medicine according to John Bell (page 100), providing a new taxonomy for human disease and triggering a change to health care practice. On page 126, Gustav Nossal reviews how an understanding of DNA processes, such as recombination, have transformed the field of immunology.

As a visual icon, and as a profound influence on our nature, the DNA molecule has permeated the imagery and art of our time, and is described by Martin Kemp (page 102) as the *Mona Lisa* of this scientific age. Given that broad impact, and revolutions that are yet to come, it is perhaps appropriate to leave the last word to an artist. Written in 1917, the poem *Heredity* by Thomas Hardy (see inset) seems to foreshadow both the essence and the fascination of the molecule that we celebrate here. □

Carina Dennis Commissioning Editor
Philip Campbell Editor, *Nature*

doi:10.1038/nature01396

Original reference: *Nature* **421**, 396 (2003).

Heredity

I am the family face;
Flesh perishes, I live on,
Projecting trait and trace
Through time to times anon,
And leaping from place to place
Over oblivion.

The years-heired feature that can
In curve and voice and eye
Despise the human span
Of durance — that is I;
The eternal thing in man,
That heeds no call to die.

Thomas Hardy
(First published in *Moments of Vision and Miscellaneous Verses*, Macmillan, 1917)

No. 4356   April 25, 1953   NATURE   737

equipment, and to Dr. G. E. R. Deacon and the captain and officers of R.R.S. *Discovery II* for their part in making the observations.

[1] Young, F. B., Gerrard, H., and Jevons, W., *Phil. Mag.*, **40**, 149 (1920).

[2] Longuet-Higgins, M. S., *Mon. Not. Roy. Astro. Soc., Geophys. Supp.*, **5**, 285 (1949).

[3] Von Arx, W. S., Woods Hole Papers in Phys. Oceanog. Meteor., **11** (3) (1950).

[4] Ekman, V. W., *Arkiv. Mat. Astron. Fysik. (Stockholm)*, **2** (11) (1905).

MOLECULAR STRUCTURE OF NUCLEIC ACIDS

A Structure for Deoxyribose Nucleic Acid

WE wish to suggest a structure for the salt of deoxyribose nucleic acid (D.N.A.). This structure has novel features which are of considerable biological interest.

A structure for nucleic acid has already been proposed by Pauling and Corey[1]. They kindly made their manuscript available to us in advance of publication. Their model consists of three intertwined chains, with the phosphates near the fibre axis, and the bases on the outside. In our opinion, this structure is unsatisfactory for two reasons : (1) We believe that the material which gives the X-ray diagrams is the salt, not the free acid. Without the acidic hydrogen atoms it is not clear what forces would hold the structure together, especially as the negatively charged phosphates near the axis will repel each other. (2) Some of the van der Waals distances appear to be too small.

Another three-chain structure has also been suggested by Fraser (in the press). In his model the phosphates are on the outside and the bases on the inside, linked together by hydrogen bonds. This structure as described is rather ill-defined, and for this reason we shall not comment on it.

We wish to put forward a radically different structure for the salt of deoxyribose nucleic acid. This structure has two helical chains each coiled round the same axis (see diagram). We have made the usual chemical assumptions, namely, that each chain consists of phosphate diester groups joining β-D-deoxyribofuranose residues with 3′,5′ linkages. The two chains (but not their bases) are related by a dyad perpendicular to the fibre axis. Both chains follow right-handed helices, but owing to the dyad the sequences of the atoms in the two chains run in opposite directions. Each chain loosely resembles Furberg's[2] model No. 1 ; that is, the bases are on the inside of the helix and the phosphates on the outside. The configuration of the sugar and the atoms near it is close to Furberg's 'standard configuration', the sugar being roughly perpendicular to the attached base. There

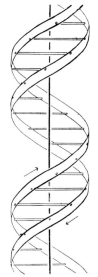

This figure is purely diagrammatic. The two ribbons symbolize the two phosphate—sugar chains, and the horizontal rods the pairs of bases holding the chains together. The vertical line marks the fibre axis

is a residue on each chain every 3·4 A. in the z-direction. We have assumed an angle of 36° between adjacent residues in the same chain, so that the structure repeats after 10 residues on each chain, that is, after 34 A. The distance of a phosphorus atom from the fibre axis is 10 A. As the phosphates are on the outside, cations have easy access to them.

The structure is an open one, and its water content is rather high. At lower water contents we would expect the bases to tilt so that the structure could become more compact.

The novel feature of the structure is the manner in which the two chains are held together by the purine and pyrimidine bases. The planes of the bases are perpendicular to the fibre axis. They are joined together in pairs, a single base from one chain being hydrogen-bonded to a single base from the other chain, so that the two lie side by side with identical z-co-ordinates. One of the pair must be a purine and the other a pyrimidine for bonding to occur. The hydrogen bonds are made as follows : purine position 1 to pyrimidine position 1 ; purine position 6 to pyrimidine position 6.

If it is assumed that the bases only occur in the structure in the most plausible tautomeric forms (that is, with the keto rather than the enol configurations) it is found that only specific pairs of bases can bond together. These pairs are : adenine (purine) with thymine (pyrimidine), and guanine (purine) with cytosine (pyrimidine).

In other words, if an adenine forms one member of a pair, on either chain, then on these assumptions the other member must be thymine ; similarly for guanine and cytosine. The sequence of bases on a single chain does not appear to be restricted in any way. However, if only specific pairs of bases can be formed, it follows that if the sequence of bases on one chain is given, then the sequence on the other chain is automatically determined.

It has been found experimentally[3,4] that the ratio of the amounts of adenine to thymine, and the ratio of guanine to cytosine, are always very close to unity for deoxyribose nucleic acid.

It is probably impossible to build this structure with a ribose sugar in place of the deoxyribose, as the extra oxygen atom would make too close a van der Waals contact.

The previously published X-ray data[5,6] on deoxyribose nucleic acid are insufficient for a rigorous test of our structure. So far as we can tell, it is roughly compatible with the experimental data, but it must be regarded as unproved until it has been checked against more exact results. Some of these are given in the following communications. We were not aware of the details of the results presented there when we devised our structure, which rests mainly though not entirely on published experimental data and stereochemical arguments.

It has not escaped our notice that the specific pairing we have postulated immediately suggests a possible copying mechanism for the genetic material.

Full details of the structure, including the conditions assumed in building it, together with a set of co-ordinates for the atoms, will be published elsewhere.

We are much indebted to Dr. Jerry Donohue for constant advice and criticism, especially on interatomic distances. We have also been stimulated by a knowledge of the general nature of the unpublished experimental results and ideas of Dr. M. H. F. Wilkins, Dr. R. E. Franklin and their co-workers at

738   NATURE   April 25, 1953   VOL. 171

King's College, London. One of us (J. D. W.) has been aided by a fellowship from the National Foundation for Infantile Paralysis.

J. D. WATSON
F. H. C. CRICK
Medical Research Council Unit for the
Study of the Molecular Structure of
Biological Systems,
Cavendish Laboratory, Cambridge.
April 2.

[1] Pauling, L., and Corey, R. B., *Nature*, **171**, 346 (1953); *Proc. U.S. Nat. Acad. Sci.*, **39**, 84 (1953).
[2] Furberg, S., *Acta Chem. Scand.*, **6**, 634 (1952).
[3] Chargaff, E., for references see Zamenhof, S., Brawerman, G., and Chargaff, E., *Biochim. et Biophys. Acta*, **9**, 402 (1952).
[4] Wyatt, G. R., *J. Gen. Physiol.*, **36**, 201 (1952).
[5] Astbury, W. T., Symp. Soc. Exp. Biol. 1, Nucleic Acid, 66 (Camb. Univ. Press, 1947).
[6] Wilkins, M. H. F., and Randall, J. T., *Biochim. et Biophys. Acta*, **10**, 192 (1953).

Molecular Structure of Deoxypentose Nucleic Acids

WHILE the biological properties of deoxypentose nucleic acid suggest a molecular structure containing great complexity, X-ray diffraction studies described here (cf. Astbury[1]) show the basic molecular configuration has great simplicity. The purpose of this communication is to describe, in a preliminary way, some of the experimental evidence for the poly-nucleotide chain configuration being helical, and existing in this form when in the natural state. A fuller account of the work will be published shortly.

The structure of deoxypentose nucleic acid is the same in all species (although the nitrogen base ratios alter considerably) in nucleoprotein, extracted or in cells, and in purified nucleate. The same linear group of polynucleotide chains may pack together parallel in different ways to give crystalline[1-3], semi-crystalline or paracrystalline material. In all cases the X-ray diffraction photograph consists of two regions, one determined largely by the regular spacing of nucleotides along the chain, and the other by the longer spacings of the chain configuration. The sequence of different nitrogen bases along the chain is not made visible.

Oriented paracrystalline deoxypentose nucleic acid ('structure B' in the following communication by Franklin and Gosling) gives a fibre diagram as shown in Fig. 1 (cf. ref. 4). Astbury suggested that the strong 3·4-A. reflexion corresponded to the inter-nucleotide repeat along the fibre axis. The ∼ 34 A. layer lines, however, are not due to a repeat of a polynucleotide composition, but to the chain configuration repeat, which causes strong diffraction as the nucleotide chains have higher density than the interstitial water. The absence of reflexions on or near the meridian immediately suggests a helical structure with axis parallel to fibre length.

Diffraction by Helices

It may be shown[5] (also Stokes, unpublished) that the intensity distribution in the diffraction pattern of a series of points equally spaced along a helix is given by the squares of Bessel functions. A uniform continuous helix gives a series of layer lines of spacing corresponding to the helix pitch, the intensity distribution along the nth layer line being proportional to the square of J_n, the nth order Bessel function. A straight line may be drawn approximately through

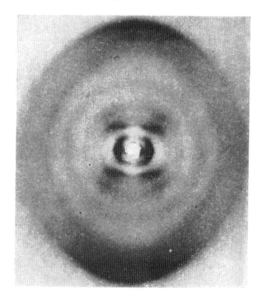

Fig. 1. Fibre diagram of deoxypentose nucleic acid from *B. coli*. Fibre axis vertical

the innermost maxima of each Bessel function and the origin. The angle this line makes with the equator is roughly equal to the angle between an element of the helix and the helix axis. If a unit repeats n times along the helix there will be a meridional reflexion ($J_0{}^2$) on the nth layer line. The helical configuration produces side-bands on this fundamental frequency, the effect[5] being to reproduce the intensity distribution about the origin around the new origin, on the nth layer line, corresponding to C in Fig. 2.

We will now briefly analyse in physical terms some of the effects of the shape and size of the repeat unit or nucleotide on the diffraction pattern. First, if the nucleotide consists of a unit having circular symmetry about an axis parallel to the helix axis, the whole diffraction pattern is modified by the form factor of the nucleotide. Second, if the nucleotide consists of a series of points on a radius at right-angles to the helix axis, the phases of radiation scattered by the helices of different diameter passing through each point are the same. Summation of the corresponding Bessel functions gives reinforcement for the inner-

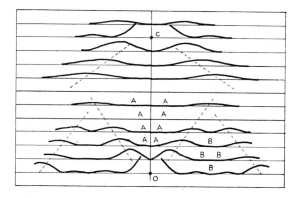

Fig. 2. Diffraction pattern of system of helices corresponding to structure of deoxypentose nucleic acid. The squares of Bessel functions are plotted about 0 on the equator and on the first, second, third and fifth layer lines for half of the nucleotide mass at 20 A. diameter and remainder distributed along a radius, the mass at a given radius being proportional to the radius. About C on the tenth layer line similar functions are plotted for an outer diameter of 12 A.

No. 4356 **April 25, 1953** NATURE 739

most maxima and, in general, owing to phase difference, cancellation of all other maxima. Such a system of helices (corresponding to a spiral staircase with the core removed) diffracts mainly over a limited angular range, behaving, in fact, like a periodic arrangement of flat plates inclined at a fixed angle to the axis. Third, if the nucleotide is extended as an arc of a circle in a plane at right-angles to the helix axis, and with centre at the axis, the intensity of the system of Bessel function layer-line streaks emanating from the origin is modified owing to the phase differences of radiation from the helices drawn through each point on the nucleotide. The form factor is that of the series of points in which the helices intersect a plane drawn through the helix axis. This part of the diffraction pattern is then repeated as a whole with origin at C (Fig. 2). Hence this aspect of nucleotide shape affects the central and peripheral regions of each layer line differently.

Interpretation of the X-Ray Photograph

It must first be decided whether the structure consists of essentially one helix giving an intensity distribution along the layer lines corresponding to $J_1, J_2, J_3 \ldots$, or two similar co-axial helices of twice the above size and relatively displaced along the axis a distance equal to half the pitch giving $J_2, J_4, J_6 \ldots$, or three helices, etc. Examination of the width of the layer-line streaks suggests the intensities correspond more closely to $J_1{}^2, J_2{}^2, J_3{}^2$ than to $J_2{}^2, J_4{}^2, J_6{}^2 \ldots$ Hence the dominant helix has a pitch of ~ 34 A., and, from the angle of the helix, its diameter is found to be ~ 20 A. The strong equatorial reflexion at ~ 17 A. suggests that the helices have a maximum diameter of ~ 20 A. and are hexagonally packed with little interpenetration. Apart from the width of the Bessel function streaks, the possibility of the helices having twice the above dimensions is also made unlikely by the absence of an equatorial reflexion at ~ 34 A. To obtain a reasonable number of nucleotides per unit volume in the fibre, two or three intertwined coaxial helices are required, there being ten nucleotides on one turn of each helix.

The absence of reflexions on or near the meridian (an empty region AAA on Fig. 2) is a direct consequence of the helical structure. On the photograph there is also a relatively empty region on and near the equator, corresponding to region BBB on Fig. 2. As discussed above, this absence of secondary Bessel function maxima can be produced by a radial distribution of the nucleotide shape. To make the layer-line streaks sufficiently narrow, it is necessary to place a large fraction of the nucleotide mass at ~ 20 A. diameter. In Fig. 2 the squares of Bessel functions are plotted for half the mass at 20 A. diameter, and the rest distributed along a radius, the mass at a given radius being proportional to the radius.

On the zero layer line there appears to be a marked $J_{10}{}^2$, and on the first, second and third layer lines, $J_9{}^2 + J_{11}{}^2, J_8{}^2 + J_{12}{}^2$, etc., respectively. This means that, in projection on a plane at right-angles to the fibre axis, the outer part of the nucleotide is relatively concentrated, giving rise to high-density regions spaced c. 6 A. apart around the circumference of a circle of 20 A. diameter. On the fifth layer line two J_5 functions overlap and produce a strong reflexion. On the sixth, seventh and eighth layer lines the maxima correspond to a helix of diameter ~ 12 A. Apparently it is only the central region of the helix structure which is well divided by the 3·4-A. spacing, the outer parts of the nucleotide overlapping to form a continuous helix. This suggests the presence of nitrogen bases arranged like a pile of pennies[1] in the central regions of the helical system.

There is a marked absence of reflexions on layer lines beyond the tenth. Disorientation in the specimen will cause more extension along the layer lines of the Bessel function streaks on the eleventh, twelfth and thirteenth layer lines than on the ninth, eighth and seventh. For this reason the reflexions on the higher-order layer lines will be less readily visible. The form factor of the nucleotide is also probably causing diminution of intensity in this region. Tilting of the nitrogen bases could have such an effect.

Reflexions on the equator are rather inadequate for determination of the radial distribution of density in the helical system. There are, however, indications that a high-density shell, as suggested above, occurs at diameter ~ 20 A.

The material is apparently not completely para-crystalline, as sharp spots appear in the central region of the second layer line, indicating a partial degree of order of the helical units relative to one another in the direction of the helix axis. Photographs similar to Fig. 1 have been obtained from sodium nucleate from calf and pig thymus, wheat germ, herring sperm, human tissue and T_2 bacteriophage. The most marked correspondence with Fig. 2 is shown by the exceptional photograph obtained by our colleagues, R. E. Franklin and R. G. Gosling, from calf thymus deoxypentose nucleate (see following communication).

It must be stressed that some of the above discussion is not without ambiguity, but in general there appears to be reasonable agreement between the experimental data and the kind of model described by Watson and Crick (see also preceding communication).

It is interesting to note that if there are ten phosphate groups arranged on each helix of diameter 20 A. and pitch 34 A., the phosphate ester backbone chain is in an almost fully extended state. Hence, when sodium nucleate fibres are stretched[3], the helix is evidently extended in length like a spiral spring in tension.

Structure *in vivo*

The biological significance of a two-chain nucleic acid unit has been noted (see preceding communication). The evidence that the helical structure discussed above does, in fact, exist in intact biological systems is briefly as follows :

Sperm heads. It may be shown that the intensity of the X-ray spectra from crystalline sperm heads is determined by the helical form-function in Fig. 2. Centrifuged trout semen give the same pattern as the dried and rehydrated or washed sperm heads used previously[6]. The sperm head fibre diagram is also given by extracted or synthetic[1] nucleoprotamine or extracted calf thymus nucleohistone.

Bacteriophage. Centrifuged wet pellets of T_2 phage photographed with X-rays while sealed in a cell with mica windows give a diffraction pattern containing the main features of paracrystalline sodium nucleate as distinct from that of crystalline nucleoprotein. This confirms current ideas of phage structure.

Transforming principle (in collaboration with H. Ephrussi-Taylor). Active deoxypentose nucleate allowed to dry at ~ 60 per cent humidity has the same crystalline structure as certain samples[3] of sodium thymonucleate.

740   N A T U R E   April 25, 1953 VOL. 171

We wish to thank Prof. J. T. Randall for encouragement ; Profs. E. Chargaff, R. Signer, J. A. V. Butler and Drs. J. D. Watson, J. D. Smith, L. Hamilton, J. C. White and G. R. Wyatt for supplying material without which this work would have been impossible ; also Drs. J. D. Watson and Mr. F. H. C. Crick for stimulation, and our colleagues R. E. Franklin, R. G. Gosling, G. L. Brown and W. E. Seeds for discussion. One of us (H. R. W.) wishes to acknowledge the award of a University of Wales Fellowship.

M. H. F. WILKINS
Medical Research Council Biophysics
Research Unit,

A. R. STOKES
H. R. WILSON
Wheatstone Physics Laboratory,
King's College, London.
April 2.

[1] Astbury, W. T., Symp. Soc. Exp. Biol., 1, Nucleic Acid (Cambridge Univ. Press, 1947).
[2] Riley, D. P., and Oster, G., Biochim. et Biophys. Acta, 7, 526 (1951).
[3] Wilkins, M. H. F., Gosling, R. G., and Seeds, W. E., Nature, 167, 759 (1951).
[4] Astbury, W. T., and Bell, F. O., Cold Spring Harb. Symp. Quant. Biol., 6, 109 (1938).
[5] Cochran, W., Crick, F. H. C., and Vand, V., Acta Cryst., 5, 581 (1952).
[6] Wilkins, M. H. F., and Randall, J. T., Biochim. et Biophys. Acta, 10, 192 (1953).

Molecular Configuration in Sodium Thymonucleate

SODIUM thymonucleate fibres give two distinct types of X-ray diagram. The first corresponds to a crystalline form, structure A, obtained at about 75 per cent relative humidity ; a study of this is described in detail elsewhere[1]. At higher humidities a different structure, structure B, showing a lower degree of order, appears and persists over a wide range of ambient humidity. The change from A to B is reversible. The water content of structure B fibres which undergo this reversible change may vary from 40–50 per cent to several hundred per cent of the dry weight. Moreover, some fibres never show structure A, and in these structure B can be obtained with an even lower water content.

The X-ray diagram of structure B (see photograph) shows in striking manner the features characteristic of helical structures, first worked out in this laboratory by Stokes (unpublished) and by Crick, Cochran and Vand[2]. Stokes and Wilkins were the first to propose such structures for nucleic acid as a result of direct studies of nucleic acid fibres, although a helical structure had been previously suggested by Furberg (thesis, London, 1949) on the basis of X-ray studies of nucleosides and nucleotides.

While the X-ray evidence cannot, at present, be taken as direct proof that the structure is helical, other considerations discussed below make the existence of a helical structure highly probable.

Structure B is derived from the crystalline structure A when the sodium thymonucleate fibres take up quantities of water in excess of about 40 per cent of their weight. The change is accompanied by an increase of about 30 per cent in the length of the fibre, and by a substantial re-arrangement of the molecule. It therefore seems reasonable to suppose that in structure B the structural units of sodium thymonucleate (molecules on groups of molecules) are relatively free from the influence of neighbouring

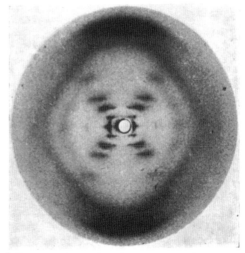

Sodium deoxyribose nucleate from calf thymus. Structure B

molecules, each unit being shielded by a sheath of water. Each unit is then free to take up its least-energy configuration independently of its neighbours and, in view of the nature of the long-chain molecules involved, it is highly likely that the general form will be helical[3]. If we adopt the hypothesis of a helical structure, it is immediately possible, from the X-ray diagram of structure B, to make certain deductions as to the nature and dimensions of the helix.

The innermost maxima on the first, second, third and fifth layer lines lie approximately on straight lines radiating from the origin. For a smooth single-strand helix the structure factor on the nth layer line is given by :

$$F_n = J_n(2\pi rR) \exp i\, n(\psi + \tfrac{1}{2}\pi),$$

where $J_n(u)$ is the nth-order Bessel function of u, r is the radius of the helix, and R and ψ are the radial and azimuthal co-ordinates in reciprocal space[2] ; this expression leads to an approximately linear array of intensity maxima of the type observed, corresponding to the first maxima in the functions J_1, J_2, J_3, etc.

If, instead of a smooth helix, we consider a series of residues equally spaced along the helix, the transform in the general case treated by Crick, Cochran and Vand is more complicated. But if there is a whole number, m, of residues per turn, the form of the transform is as for a smooth helix with the addition, only, of the same pattern repeated with its origin at heights mc^*, $2mc^*$ . . . etc. (c is the fibre-axis period).

In the present case the fibre-axis period is 34 A. and the very strong reflexion at 3·4 A. lies on the tenth layer line. Moreover, lines of maxima radiating from the 3·4-A. reflexion as from the origin are visible on the fifth and lower layer lines, having a J_5 maximum coincident with that of the origin series on the fifth layer line. (The strong outer streaks which apparently radiate from the 3·4-A. maximum are not, however, so easily explained.) This suggests strongly that there are exactly 10 residues per turn of the helix. If this is so, then from a measurement of R_n the position of the first maximum on the nth layer line (for $n \leqslant 5$), the radius of the helix, can be obtained. In the present instance, measurements of R_1, R_2, R_3 and R_5 all lead to values of r of about 10 A.

No. 4356   **April 25, 1953**   N A T U R E   741

Since this linear array of maxima is one of the strongest features of the X-ray diagram, we must conclude that a crystallographically important part of the molecule lies on a helix of this diameter. This can only be the phosphate groups or phosphorus atoms.

If ten phosphorus atoms lie on one turn of a helix of radius 10 A., the distance between neighbouring phosphorus atoms in a molecule is 7·1 A. This corresponds to the P . . . P distance in a fully extended molecule, and therefore provides a further indication that the phosphates lie on the outside of the structural unit.

Thus, our conclusions differ from those of Pauling and Corey[4], who proposed for the nucleic acids a helical structure in which the phosphate groups form a dense core.

We must now consider briefly the equatorial reflexions. For a single helix the series of equatorial maxima should correspond to the maxima in $J_0(2\pi rR)$. The maxima on our photograph do not, however, fit this function for the value of r deduced above. There is a very strong reflexion at about 24 A. and then only a faint sharp reflexion at 9·0 A. and two diffuse bands around 5·5 A. and 4·0 A. This lack of agreement is, however, to be expected, for we know that the helix so far considered can only be the most important member of a series of coaxial helices of different radii ; the non-phosphate parts of the molecule will lie on inner co-axial helices, and it can be shown that, whereas these will not appreciably influence the innermost maxima on the layer lines, they may have the effect of destroying or shifting both the equatorial maxima and the outer maxima on other layer lines.

Thus, if the structure is helical, we find that the phosphate groups or phosphorus atoms lie on a helix of diameter about 20 A., and the sugar and base groups must accordingly be turned inwards towards the helical axis.

Considerations of density show, however, that a cylindrical repeat unit of height 34 A. and diameter 20 A. must contain many more than ten nucleotides.

Since structure B often exists in fibres with low water content, it seems that the density of the helical unit cannot differ greatly from that of dry sodium thymonucleate, 1·63 gm./cm.[3] [1,5], the water in fibres of high water-content being situated outside the structural unit. On this basis we find that a cylinder of radius 10 A. and height 34 A. would contain thirty-two nucleotides. However, there might possibly be some slight inter-penetration of the cylindrical units in the dry state making their effective radius rather less. It is therefore difficult to decide, on the basis of density measurements alone, whether one repeating unit contains ten nucleotides on each of two or on each of three co-axial molecules. (If the effective radius were 8 A. the cylinder would contain twenty nucleotides.) Two other arguments, however, make it highly probable that there are only two co-axial molecules.

First, a study of the Patterson function of structure A, using superposition methods, has indicated[6] that there are only two chains passing through a primitive unit cell in this structure. Since the $A \rightleftharpoons B$ transformation is readily reversible, it seems very unlikely that the molecules would be grouped in threes in structure B. Secondly, from measurements on the X-ray diagram of structure B it can readily be shown that, whether the number of chains per unit is two or three, the chains are not equally spaced along the

fibre axis. For example, three equally spaced chains would mean that the nth layer line depended on J_{3n}, and would lead to a helix of diameter about 60 A. This is many times larger than the primitive unit cell in structure A, and absurdly large in relation to the dimensions of nucleotides. Three unequally spaced chains, on the other hand, would be crystallographically non-equivalent, and this, again, seems unlikely. It therefore seems probable that there are only two co-axial molecules and that these are unequally spaced along the fibre axis.

Thus, while we do not attempt to offer a complete interpretation of the fibre-diagram of structure B, we may state the following conclusions. The structure is probably helical. The phosphate groups lie on the outside of the structural unit, on a helix of diameter about 20 A. The structural unit probably consists of two co-axial molecules which are not equally spaced along the fibre axis, their mutual displacement being such as to account for the variation of observed intensities of the innermost maxima on the layer lines ; if one molecule is displaced from the other by about three-eighths of the fibre-axis period, this would account for the absence of the fourth layer line maxima and the weakness of the sixth. Thus our general ideas are not inconsistent with the model proposed by Watson and Crick in the preceding communication.

The conclusion that the phosphate groups lie on the outside of the structural unit has been reached previously by quite other reasoning[1]. Two principal lines of argument were invoked. The first derives from the work of Gulland and his collaborators[7], who showed that even in aqueous solution the —CO and —NH₂ groups of the bases are inaccessible and cannot be titrated, whereas the phosphate groups are fully accessible. The second is based on our own observations[1] on the way in which the structural units in structures A and B are progressively separated by an excess of water, the process being a continuous one which leads to the formation first of a gel and ultimately to a solution. The hygroscopic part of the molecule may be presumed to lie in the phosphate groups ($(C_2H_5O)_2PO_2Na$ and $(C_3H_7O)_2PO_2Na$ are highly hygroscopic[8]), and the simplest explanation of the above process is that these groups lie on the outside of the structural units. Moreover, the ready availability of the phosphate groups for interaction with proteins can most easily be explained in this way.

We are grateful to Prof. J. T. Randall for his interest and to Drs. F. H. C. Crick, A. R. Stokes and M. H. F. Wilkins for discussion. One of us (R. E. F.) acknowledges the award of a Turner and Newall Fellowship.

ROSALIND E. FRANKLIN*
R. G. GOSLING

Wheatstone Physics Laboratory,
    King's College, London.
        April 2.

* Now at Birkbeck College Research Laboratories, 21 Torrington Square, London, W.C.1.

[1] Franklin, R. E., and Gosling, R. G. (in the press).

[2] Cochran, W., Crick, F. H. C., and Vand, V., *Acta Cryst.*, **5**, 501 (1952).

[3] Pauling, L., Corey, R. B., and Bransom, H. R., *Proc. U.S. Nat. Acad. Sci.*, **37**, 205 (1951).

[4] Pauling, L., and Corey, R. B., *Proc. U.S. Nat. Acad. Sci.*, **39**, 84 (1953).

[5] Astbury, W. T., Cold Spring Harbor Symp. on Quant. Biol., **12**, 56 (1947).

[6] Franklin, R. E., and Gosling, R. G. (to be published).

[7] Gulland, J. M., and Jordan, D. O., Cold Spring Harbor Symp. on Quant. Biol., **12**, 5 (1947).

[8] Drushel, W. A., and Felty, A. R., *Chem. Zent.*, **89**, 1016 (1918).

Quiet debut for the double helix

Robert Olby

Department of the History and Philosophy of Science, 1017 Cathedral of Learning, University of Pittsburgh, Pittsburgh, Pennsylvania 15260, USA
(e-mail: olbyr+@pitt.edu)

Past discoveries usually become aggrandized in retrospect, especially at jubilee celebrations, and the double helix is no exception. The historical record reveals a muted response by the scientific community to the proposal of this structure in 1953. Indeed, it was only when the outlines appeared of a mechanism for DNA's involvement in protein synthesis that the biochemical community began to take a serious interest in the structure.

"... we may expect genetic chemistry to become in time an integrating core for cellular biochemistry." Robert Sinsheimer, in a lecture delivered at the California Institute of Technology, 1956 (published in ref. 1, p. 1128).

To recall the year 1953 is to visit — and for some of us to revisit — another world, when *Nature* did not use the abbreviation DNA for deoxyribonucleic acid. In June that year, Elizabeth II, Queen of the United Kingdom, was crowned amidst much pomp and ceremony. In March, British scientists prepared to construct an atomic power station by the Calder River. Two months later, Mount Everest was conquered. At the University of London my biochemistry teacher enthused about Frederick Sanger's success in the first sequencing of the units of a protein, insulin. But deoxyribonucleic acid (DNA) was not even mentioned. Yet in 1953 *Nature* published seven papers on the structure and function of DNA[2-8], but only one national British newspaper — the *News Chronicle* — referred to the double helix[9] (see facsimile below).

Reception to the double helix

Fifty years on it is hard to believe the double helix had such a lukewarm reception. But turn to *Nature* and to *Science* in the 1950s and what do we find? Figure 1 records the number of papers in *Nature* reporting on any aspects of DNA, and of these the number that mention the Watson–Crick model or cite any of the 1953 papers on DNA structure. Through the decade *Nature*'s volumes increased in size, and in 1960 the number of volumes published per year was doubled. This increase was accompanied by an increase in the number of papers on some aspect of DNA, but references to the double helix did not increase. The pattern of citation in *Science* is similar.

At the time the structure of DNA was discovered, there was already a considerable ongoing programme

Ritchie Calder's report on the discovery of the structure of DNA on page 1 of the *News Chronicle*, 15 May 1953.

of research on DNA (see time line in Box 1). These studies include the physical properties of DNA, methods of extraction, and whether the content and composition of DNA is the same for all the cells of the same organism. Also discussed were the damaging effects of ultraviolet light and ionizing radiation on DNA, and differing views over the involvement of nucleic acids in protein synthesis.

Researchers working on DNA at that time were principally biochemists and physical chemists, and their institutional locations and funding were chiefly medically related. Their interests and means of support related to two main concerns of the time — the action of 'mutagens' (agents that cause mutations in DNA), a subject important to the international debate on the effects of ionizing radiation and radioactive materials (see accompanying article by Friedberg, page 122), and the nature of protein synthesis, of great interest to biochemists in the light of its importance in growth and nutrition, in addition to cancer research.

In the light of the muted reception of the structure, let us take a different angle and ask what justification was there in the 1950s for giving the DNA double helix more than passing attention? At the time, most scientists reading *Nature* viewed DNA as a 'conjugated protein', owing to its association with protein; it was important as such, but not in its own right. This was despite the remarkable work of Oswald Avery, Colin MacLeod and Maclyn McCarty in 1944 (ref. 10; and see accompanying article by McCarty, page 92), followed by Al Hershey and Martha Chase's demonstration in 1952 (ref. 11) that most of the material entering a bacterium from an infecting bacterial virus is nucleic acid not protein. These studies made DNA look very much like the hereditary material.

Connecting structure to function

More information was needed to convince the scientific community. What was there about the chemistry of DNA to justify its role in inheritance? An answer came with the structure put forward by Watson and Crick. Chief among its "novel features" of "considerable biological interest"[2], Watson and Crick described the pairing of the bases, where adenine forms hydrogen bonds with thymine, and guanine with cytosine. This pairing, they wrote, "immediately suggests a possible copying mechanism for the genetic material."[2] Expanding on this in a subsequent paper appearing in *Nature* a month later, they wrote of DNA: "Until now, however, no evidence has been

No one suggests these groupings can yet be arranged artificially. Discovering how these chemical "cards" are shuffled and paired will keep the scientists busy for the next 50 years.

presented to show how it might carry out the essential operation required of a genetic material, that of exact self-duplication."[5]

With these words Watson and Crick claimed their priority on a mechanism for DNA replication, but admitted there were problems with their scheme: how do the chains unwind and separate "without everything getting tangled"[5]? What is the exact mechanism by which gene duplication occurs? How does the genetic material "exert a highly specific influence on the cell"[12] when the sequence of bases assumed to encode the specificity is on the inside of the helical molecule?

The 'unwinding problem' dominated much of the early discussions that followed the discovery of the DNA structure. In 1953, Watson and Crick admitted it was "formidable"[12], but support for their structure came in 1958, when Matthew Meselson and Franklin Stahl proved the semi-conservative nature of DNA replication[13]: each of the two new daughter DNA molecules formed during DNA replication consists of one strand from the original parent molecule and a new strand synthesized from the parent strand, which served as a template. This confirmed Watson and Crick's theoretical prediction from the structure that replication would proceed in a semi-conservative manner. Later that same year, Arthur Kornberg announced the partial purification of an enzyme that catalyses DNA synthesis later called DNA polymerase[14]. This first linked enzymology to the double helix, for not long thereafter Kornberg provided biochemical evidence that DNA polymerase synthesizes new strands from opposite directions of the two chains of the molecule[15].

In 1957, Crick defined biological 'information' as the sequence of the bases in the nucleic acids and of the amino acids in proteins, and proposed the now famous 'central dogma' according to which information so defined flows between the nucleic acids and proteins only in one direction — from the former to the latter[16]. Just four years later, Marshall Nirenberg and Heinrich Matthaei successfully synthesized a polypeptide constituted of only one kind of amino acid (phenylalanine) using an RNA composed only of one kind of base (uracil). They concluded that "one or more [of these RNA bases] appear to be the code for phenylalanine."[17] Meanwhile, Crick, Sydney Brenner and Leslie Barnett had been using genetic analysis to investigate mutagenesis. This led them to the important concept of a form of mutation in which there is a 'frame shift' in the sequence of the bases in DNA, from which they went on to infer that the genetic message is composed of single or multiple triplets of bases, and that the message is read starting at a fixed point and proceeds always in the same

Six of the Nobel winners of 1962 display their diplomas after formal ceremonies in Stockholm's Concert Hall. From left to right: Maurice Wilkins (Medicine), Max Perutz (Chemistry), Francis Crick (Medicine), John Steinbeck (Literature), James Watson (Medicine) and John Kendrew (Chemistry).

Figure 1 Papers published in *Nature* referring to DNA and the extent of their reference to the double helix 1950–1960.

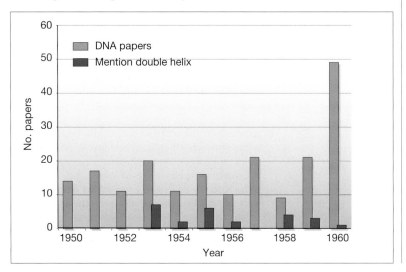

Box 1
Time line of the discovery of the structure of DNA

1869   Fritz Miescher discovers that the nuclei of pus cells contain an acidic substance to which he gave the name 'nuclein'. Later he finds that nuclein is composed of a protein and a compound to which the name nucleic acid, and subsequently DNA, will be given.

1919   Phoebus Aaron Levene proposes the 'tetranucleotide' structure of DNA, whereby the four bases of DNA were arranged one after another in a set of four.

1928   Frederick Griffith finds that a substance in heat-killed bacteria can cause heritable changes in the live bacteria alongside them. He calls the phenomenon 'transformation'.

1938   Rudolf Signer, Torbjorn Caspersson and Einer Hammarsten find molecular weights for DNA between 500,000 and 1,000,000 daltons. Levene's tetranucleotide must be a polytetranucleotide.

1944   Oswald Avery, Colin MacLeod and Maclyn McCarty establish the chemical identity of Griffith's transforming principle as DNA, and they suggest that it may function as the genetic material.

1949   Erwin Chargaff reports that DNA base composition varies from one species to another, yet the ratio between the quantities of the two purine bases, adenine and guanine, and that between the quantities of the two pyrimidine bases, thymine and cytosine, remains about the same, namely one to one.

1949   Roger and Colette Vendrely, together with André Boivin find half as much DNA in the nuclei of sex cells as they find in the body cells, thus paralleling the reduction in the number of chromosomes, making DNA look like the genetic material.

1951   Rosalind Franklin distinguishes two forms of DNA, the paracrystalline B form and the crystalline A form.

1952   Al Hershey and Martha Chase find that DNA but scarcely any protein from an infecting bacterial virus enters the bacterial cell and can be recovered from the progeny virus particles.

1952   Rosalind Franklin and Raymond Gosling produce a magnificent X-ray diffraction pattern of the B form of DNA.

1953   James Watson and Francis Crick, Rosalind Franklin and Raymond Gosling, Maurice Wilkins, W. E. Seeds, Alec Stokes and Herbert Wilson, and Bertil Jacobson all publish on the structure of DNA[2-8].

1954   George Gamow suggests a DNA code for the synthesis of proteins.

1955   Seymour Benzer analyses the fine structure of the genetic material of a bacterial virus at a level close to the distances that separate the individual bases along the DNA chain.

1957   Francis Crick proposes 'the sequence hypothesis' and 'the central dogma'.

1958   Matthew Meselson and Franklin Stahl demonstrate the semi-conservative replication of DNA.

1959   Arthur Kornberg and colleagues isolate the enzyme DNA polymerase.

1961   Marshall Nirenberg and Johann Heinrich Matthaei show that a sequence of nucleotide can encode a particular amino acid, laying the foundations for deciphering the genetic code.

1962   The Nobel prize in medicine is awarded to James Watson, Francis Crick and Maurice Wilkins.

direction[18]. Thus was the stage set for the subsequent unravelling of the entire genetic code.

From a muted reception in 1953 to accelerating momentum towards the end of the decade, one is tempted to infer that the DNA double helix was not taken seriously until a mechanism for its involvement in protein synthesis began to take shape. There was, to be sure, a small band of scientists who from the start either built their careers upon the implications of the structure (such as Meselson and Alexander Rich) or redirected their research to follow it up (including Seymour Benzer and Sydney Brenner). However, many scientists, notably Erwin Chargaff and

Alexander Dounce, did not refer to the structure in their scientific papers in the mid-fifties, even though it was clearly relevant and presumably known to them. Such omissions suggest that some biochemists had their own agendas, and the double helix was not at first seen as an aid to their work.

Biochemists debate protein synthesis

Biochemists' reservations about the double helix stemmed in part from the fact that evidential support for it in 1953 was far from strong. Watson and Crick themselves admitted that it "could in no sense be considered proved", although it was "most promising"[19]. In part the biochemists' coolness owed much to the debates among them over the mechanism of protein synthesis. The paper by Peter Campbell and Thomas Work, published in *Nature* on 6 June 1953, portrayed this debate vividly. They identified two contrasting theories under discussion on how proteins are made: first, the peptide theory (also known as the multi-enzyme theory), where proteins are made by "stepwise coupling of many small peptide units"; and second, the template theory, involving "synthesis on templates, each template being specific for a single protein structure and probably identifiable as a gene."[20]

The peptide model was, for a very long time, supported by many prominent biochemists, including Joseph Fruton. The conviction behind it was the power of enzymes to both synthesize and break down their substrates, with a high degree of specificity attributed to both actions. Synthesis was proposed to involve the formation of a succession of peptides, ultimately yielding the protein molecule, and enzymes synthesize only those peptide bonds that they also hydrolyse. But the problem with this theory was that, except for a very few special cases, the alleged peptides constituting the intermediaries in protein synthesis could neither be detected in the cell nor incorporated into the protein being synthesized. Amino acids, however, could be incorporated, indicating they were the building blocks of proteins.

The second model of protein synthesis, which assumed synthesis on a template, had been advocated by Dounce in 1952. He pictured polypeptide chains being laid down on RNA molecules, and the RNA sequence determining the sequence of amino acids incorporated (on a one-to-one basis). Thus, DNA in the nucleus would control the order of bases in the RNA[21].

After weighing up the merits and difficulties of Dounce's scheme, Campbell and Work voiced their distaste for the genetic control of protein synthesis, remarking in 1953 that: "...the gene is essentially an abstract idea and it may be a mistake to try to clothe this idea in a coat of nucleic acid or protein... if we must have a gene it should have a negative rather than a positive function so far as protein synthesis is concerned."[20] Only three years later, however, Robert Sinsheimer concluded a lecture at the California Institute of Technology with the following words: "The gene, once a formal abstraction, has begun to condense, to assume form and structure and defined activity."[1]

But those three years were a scene of pronounced change. By January 1957, when Fruton revised the second edition of his widely used textbook *General Biochemistry*, his remarks on the peptide theory were cautious and were followed by a discussion of the role of RNA on which, he noted, there have been "stimulating speculations about the role of nucleic acids as 'templates' in protein synthesis."[22] Earlier in the book he devoted a paragraph to the double helix, describing it as an 'ingenious speculation'. The only diagram was of the base pair adenine–thymine, rather than the helical model of the structure.

Kornberg had shown in 1957 that DNA replication follows the rules of base pairing, whereby DNA polymerase adds a base to the newly synthesized strand that is complementary to the opposing base in the template strand (A is always opposite T, and C always opposite G). But his interest in the subject had not been stimulated by Watson and Crick's discovery. Rather, in 1953 he was preoccupied with how coenzymes (non-protein compounds needed for enzyme activity) are synthesized from nucleotides. He was led to wonder how DNA and RNA might be made from thousands of nucleotides. "The significance of the double helix," he recalled, "did not intrude" into his work until 1956, after he had shown that a "moderately purified fraction" of what he was later to call DNA polymerase "appeared to increase the size of a DNA chain."[23,24]

Conclusion

The two once enigmatic processes — DNA replication and protein synthesis — intersected ongoing research programmes in the physical, organic and biological chemistry of the early 1950s. After the discovery of the double helix, those grappling with the problem of replication found its molecular foundation in the structure of DNA, although it took more than two decades to deduce the intricate mechanism of its operation in the cell (see accompanying article by Alberts, page 117). Those working on protein synthesis found the source of its specificity lay in the base sequence of DNA.

But why celebrate this one discovery? Why not celebrate the golden jubilee of Max Perutz's solution to the 'phase problem' for proteins in 1953, without which the subsequent discovery of the structure of myoglobin and haemoglobin would not have been possible? What about the year 2005 for celebrating the golden jubilee of Sanger's determination of the complete amino-acid sequence of a protein? Undoubtedly, the double helix has remarkable iconic value that has contributed significantly to its public visibility, something that has not been achieved by any of the protein structures (see accompanying article by Kemp, page 102). There is, too, a degree of notoriety attaching to the manner of its discovery and the characters involved that has given spice to the story, as widely publicized by James Watson's account of the discovery in *The Double Helix*, published in 1968 (ref. 25), and Brenda Maddox's recent illuminating biography of Rosalind Franklin[26]. But there is a centrality about DNA that relates to the centrality of heredity in general biology.

The silver and golden jubilees of the Queen's accession to the throne have come and gone, nuclear power stations are no longer being built in the United Kingdom, and mountaineer after mountaineer has ascended Mount Everest without a fanfare of press reports. But DNA is very much in the news — whether it be as a tool for studying evolution, a forensic test for rape, a source of genetic information or a path to designer drugs. And what better emblem or mascot is there for molecular biology than the double helix, and its spartan yet elegant representation in the original paper[2] from the pen of Odile Crick, Francis's wife, fifty years ago? ☐

doi:10.1038/nature01397

1. Sinsheimer, R. L. First steps toward a genetic chemistry. *Science* **125**, 1123–1128 (1957).
2. Watson, J. D. & Crick, F. H. C. A structure for deoxyribose nucleic acid. *Nature* **171**, 737–738 (1953).
3. Wilkins, M. H. F., Stokes, A. R. & Wilson, H. R. Molecular structure of deoxypentose nucleic acids. *Nature* **171**, 738–740 (1953).
4. Franklin, R. E. & Gosling, R. G. Molecular configuration in sodium thymonucleate. *Nature* **171**, 740–741 (1953).
5. Watson, J. D. & Crick, F. H. C. Genetical implications of the structure of deoxyribonucleic acid. *Nature* **171**, 964–967 (1953).
6. Franklin, R. E. & Gosling, R. G. Evidence for 2-chain helix in crystalline structure of sodium deoxyribonucleate. *Nature* **172**, 156–157 (1953).
7. Jacobson, B. Hydration structure of deoxyribonucleic acid and its physico-chemical properties. *Nature* **172**, 666–667 (1953).
8. Wilkins, M. H. F., Seeds, W. E., Stokes, A. R. & Wilson, H. R. Helical structure of crystalline deoxypentose nucleic acid. *Nature* **172**, 759–762 (1953).
9. Calder, R. Why you are you: nearer the secret of life. *News Chronicle* p. 1 (15 May 1953).
10. Avery, O. T. MacLeod, C. M. & McCarty, M. Studies of the chemical nature of the substance inducing transformation of pneumococcal types. Induction of transformation by a desoxyribonucleic acid fraction isolated from Pneumococcus Type III. *J. Exp. Med.* **79**, 137-158 (1944).
11. Hershey, A. D. & Chase, M. Independent functions of viral proteins and nucleic acid in growth of bacteriophage. *J. Gen. Physiol.* **36**, 39–56 (1952).
12. Watson, J. D. & Crick, F. H. C. The structure of DNA. *Cold Spring Harb. Symp. Quant. Biol.* **18**, 123-131 (1953).
13. Meselson, M. & Stahl, F. W. The replication of DNA in *Escherichia coli*. *Proc. Natl Acad. Sci. USA* **44**, 671–682 (1958).
14. Lehman, I. R., Bessmanm, M. J., Simms, E. S. & Kornberg, A. Enzymatic synthesis of deoxyribonucleic acid. I. Preparation of substrates and partial purification of an enzyme from *Escherichia coli*. *J. Biol. Chem.* **233**, 163–170 (1958).
15. Kornberg, A. Biological synthesis of deoxyribonucleic acid: an isolated enzyme catalyzes synthesis of this nucleic acid in response to directions from pre-existing DNA. *Science* **131**, 1503–1508 (1960).
16. Crick, F. H. C. On protein synthesis. *Symp. Soc. Exp. Biol.* **12**, 138–163 (1958).
17. Nirenberg, M. W. & Matthaei, J. H. The dependence of cell-free protein synthesis in *E. Coli* upon naturally occurring or synthetic polynucleotides. *Proc. Natl Acad. Sci. USA* **47**, 1558–1602 (1961).
18. Crick, F. H. C., Barnett, L., Brenner, S. & Watts-Tobin, R. J. General nature of the genetic code for proteins. *Nature* **192**, 1227–1232 (1961).
19. Crick, F. H. C. & Watson, J. D. The complementary structure of deoxyribonucleic acid. *Proc. R. Soc. Lond. A* **223**, 80–96 (1954).
20. Campbell, P. N. & Work, T. S. Biosynthesis of proteins. *Nature* **171**, 997–1001 (1953).
21. Dounce, A. Duplicating mechanisms for peptide chain and nucleic acid synthesis. *Enzymologia* **15**, 251–258 (1952).
22. Fruton, J. S. *General Biochemistry* 2nd edn (Wiley, New York 1958).
23. Kornberg, A. *For the Love of Enzymes. The Odyssey of a Biochemist* (Harvard Univ. Press, Cambridge, MA, 1989).
24. Kornberg, A. in *A Symposium on the Chemical Basis of Heredity* (eds McElroy, W. D. & Glass, B.) 605 (Johns Hopkins Press, Baltimore, 1957).
25. Watson, J. D. *The Double Helix: A Personal Account of the Discovery of the Structure of DNA* (Atheneum, New York, 1968). [Norton Critical Edition (ed. Stent, G. S.) published by Norton, New York & London, 1980.]
26. Maddox, B. *Rosalind Franklin. The Dark Lady of DNA* (Harper Collins, London 2002).

Original reference: *Nature* **421**, 402–405 (2003).

Discovering genes are made of DNA

Maclyn McCarty

The Rockefeller University, 1230 York Avenue, New York 10021, USA (e-mail: mccartm@rockefeller.edu)

Maclyn McCarty is the sole surviving member of the team that made the remarkable discovery that DNA is the material of inheritance. This preceded by a decade the discovery of the structure of DNA itself. Here he shares his personal perspective of those times and the impact of the double helix.

Editor's note — For a long time, biologists thought that 'genes', the units of inheritance, were made up of protein. In 1944, in what was arguably the defining moment for nucleic acid research, Oswald Avery, Maclyn McCarty and Colin MacLeod, at Rockefeller Institute (now University) Hospital, New York, proved that DNA was the material of inheritance, the so-called stuff of life. They showed that the heritable property of virulence from one infectious strain of pneumococcus (the bacterial agent of pneumonia) could be transferred to a noninfectious bacterium with pure DNA[1]. They further supported their conclusions by showing that this 'transforming' activity could be destroyed by the DNA-digesting enzyme DNAase[2,3].

This work first linked genetic information with DNA and provided the historical platform of modern genetics. Their discovery was greeted initially with scepticism, however, in part because many scientists believed that DNA was too simple a molecule to be the genetic material. And the fact that McCarty, Avery and MacLeod were not awarded the Nobel prize is an oversight that, to this day, still puzzles.

> *"The pivotal discovery of 20th-century biology."*
> Joshua Lederberg, Rockefeller University, 1994, referring to the discovery by McCarty, Avery and MacLeod.

At the time of our discovery and publication in 1944 (ref. 1) of the research showing that DNA is heritable, my personal view, which I shared with MacLeod, was that there was little doubt that genes are made of DNA, and that this would ultimately be accepted. I was not sure of the best approach to use in pursuing research on the subject, but suspected that clarification of the structure of DNA was necessary.

But this was not an area of research in which I had received any training. Additionally, I had planned to make my career in disease-oriented research, and knowledge of the gene did not seem likely to become applicable in this area for some years. Thus, when invited to lead my own laboratory in the Rockefeller Hospital, investigating streptococcal infection and the pathogenesis of rheumatic fever, I decided to leave Avery's laboratory for this new position in July 1946.

Rollin Hotchkiss joined Avery at this point, and together with Harriett Taylor (a recent PhD graduate in genetics who had joined the laboratory in 1945), carried out studies increasing the purity of the transforming DNA mixture by further reducing any contaminating traces of protein. Together with other investigators, they also showed that properties of the pneumococcus other than just specific polysaccharide components of its cell wall could be transferred by the DNA preparations, indicating that the purified DNA also contained other genes of the bacterium.

Our findings continued to receive little acceptance for a variety of reasons, the most significant being that the work on the composition of DNA, dating back to its first identification 75 years earlier, had concluded that DNA was too limited in diversity to carry genetic information. Even those biologists who had considered the possibility had dropped the idea, and the prevailing dogma was that if genes are composed of a known substance, it must be protein.

There were a few biologists who took a different view, the most notable being Erwin Chargaff, who changed his area of research to DNA after reading our 1944 paper[1]. His work revealed the great diversity in DNA isolated from various sources, and that

despite this diversity the amount of adenine always equalled that of thymine, and the amount of guanine that of cytosine. The latter finding was an important factor in the next significant advance in the field — the Watson–Crick determination of the double helical structure of DNA.

After the change in my research activity, I continued to give talks on our work on pneumococcal transformation and found the acceptance of the probable genetic role of DNA still to be minimal. However, I was convinced that it was only a matter of time before our results would become established.

Even though I was no longer involved in research on the subject, I continued to follow the developments as they appeared in the literature. Thus, when the papers of Watson and Crick describing the double helical structure of DNA were published in *Nature* in 1953, I certainly grasped the significance of their findings and was pleased to see such illuminating results come from a structural approach. I was not so pleased, however, that they failed to cite our work as one reason for pursuing the structure of DNA.

The concept of the double helix also hastened the silencing of those who had clung to the idea of genes as proteins. As a progressively larger body of investigators joined the study of the genetic role of DNA, there was an expanding amount of new information, starting with the resolution of the genetic code. By the end of the twentieth century, subsequent work on the mechanisms by which DNA is replicated with each cell division, is reshuffled with each generation, and is repaired when mistakes arise — the importance of which can in each case be traced back to the finding that DNA is the hereditary material — has transformed research in all areas of biology, technology and medicine. □

doi:10.1038/nature01398

1. Avery, O. T., MacLeod, C. M. & McCarty, M. Studies of the chemical nature of the substance inducing transformation of pneumococcal types. Induction of transformation by a desoxyribonucleic acid fraction isolated from Pneumococcus Type III. *J. Exp. Med.* **79,** 137–158 (1944).
2. McCarty, M. Purification and properties of desoxyribonuclease isolated from beef pancreas. *J. Gen. Physiol.* **29,** 123–139 (1946).
3. McCarty, M. & Avery, O. T. Studies of the chemical nature of the substance inducing transformation of pneumococcal types II. Effect of desoxyribonuclease on the biological activity of the transforming substance. *J. Exp. Med.* **83,** 89–96 (1946).

Original reference: *Nature* **421,** 406 (2003).

Maclyn McCarty at The Rockefeller University.

The double helix and the 'wronged heroine'

Brenda Maddox

9 Pitt Street, London W8 4NX, UK (e-mail: bmaddox@pitt.demon.co.uk)

In 1962, James Watson, Francis Crick and Maurice Wilkins received the Nobel prize for the discovery of the structure of DNA. Notably absent from the podium was Rosalind Franklin, whose X-ray photographs of DNA contributed directly to the discovery of the double helix. Franklin's premature death, combined with misogynist treatment by the male scientific establishment, cast her as a feminist icon. This myth overshadowed her intellectual strength and independence both as a scientist and as an individual.

"Science and everyday life cannot and should not be separated. Science, for me, gives a partial explanation of life. In so far as it goes, it is based on fact, experience and experiment." Rosalind Franklin, in a letter to her father, summer 1940.

In late February 1953, Rosalind Franklin, a 33-year-old physical chemist working in the biophysics unit of King's College in London, wrote in her notebooks that the structure of DNA had two chains. She had already worked out that the molecule had its phosphate groups on the outside and that DNA existed in two forms.

Two weeks later James Watson and Francis Crick, at the Cavendish Laboratory at Cambridge, built their now celebrated model of DNA as a double helix. They did it not only through brilliant intuition and a meeting of compatible minds, but also on the basis of Franklin's unpublished experimental evidence, which had reached them through irregular routes. She did not know that they had seen either her X-ray photograph (Fig. 1), showing unmistakable evidence of a helical structure, or her precise measurements of the unit cell (the smallest repeating unit), and the crystalline symmetry, of the DNA fibres.

As Watson was to write candidly, "Rosy, of course, did not directly give us her data. For that matter, no one at King's realized they were in our hands." When this admission appeared in Watson's best-selling, much-acclaimed book of the discovery, *The Double Helix*, published in 1968 (ref. 1), he was a Harvard professor and Nobel laureate (he had shared the prize for medicine and physiology in 1962, with Crick and Maurice Wilkins of King's College.) By then Franklin had died — in 1958, at the age of 37, from ovarian cancer.

Other comments dismissive of "Rosy" in Watson's book caught the attention of the emerging women's movement in the late 1960s. "Clearly Rosy had to go or be put in her place [...] Unfortunately Maurice could not see any decent way to give Rosy the boot". And, "Certainly a bad way to go out into the foulness of a [...] November night was to be told by a woman to refrain from venturing an opinion about a subject for which you were not trained."

A feminist icon

Such flamboyantly chauvinist phrases were sufficient to launch the legend of Franklin, the wronged heroine. So too was Watson's insistence on judging Franklin by her appearance rather than by her performance as a scientist. (She was, when she came to King's from the French government laboratory where she had worked from 1947 to the end of 1950, a recognized expert on the structure of coals, carbons and disordered crystals, with many publications to her credit.)

The Franklin myth has continued to grow, abetted by the fact of her tragically early death. Franklin has become a feminist icon — the Sylvia Plath of molecular biology — seen as a genius whose gifts were sacrificed to the greater glory of the male. Her failure to win the Nobel prize has been given as a prime example

Figure 1 "Her photographs are among the most beautiful X-ray photographs of any substance every taken." — J. D. Bernal, 1958. Franklin's X-ray diagram of the B form of sodium thymonucleate (DNA) fibres, published in *Nature* on 25 April 1953, shows "in striking manner the features characteristic of helical structures"[5].

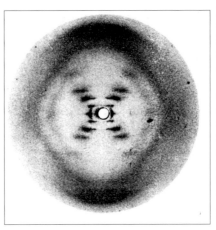

of the entrenched misogyny of the science establishment, rather than the consequence of the Nobel statute against posthumous awards.

Watson's caricature of the bad-tempered "Rosy" drew a counter-blast from her good friend, the American writer Anne Sayre, in *Rosalind Franklin and DNA*, published in 1975 (ref. 2). Sayre's book provided a much-needed corrective portrait, but was marred by a feminist bias. For example, it grossly underestimated the number of women scientists at King's in the early 1950s. Sayre maintained there was only one other than Franklin, whereas there were at least eight on the senior staff. She insisted, moreover, that women's exclusion from the King's senior common room deprived Franklin of the intellectual companionship of her colleagues. In fact, most of the scientific staff preferred to eat in the joint dining room, men and women together, and the women, in general, felt well treated at King's.

Reassessing the facts

As a biographer writing nearly three decades later and given access to Franklin's personal correspondence, I found a more attractive, capable woman than Watson had suggested, and a King's College more congenial and welcoming to women scientists than Sayre had allowed. I also found that Franklin felt singularly unhappy at King's, not so much because of her gender, but because of her class and religion: a wealthy Anglo-Jew felt out of place in a Church of England setting dominated by swirling cassocks and students studying for the priesthood. "At King's," she wrote to Sayre (albeit inaccurately), "there are neither Jews nor foreigners".

She was, in fact, so unhappy at King's that, in early 1953, getting out as fast as possible was far more important to her than finishing her work on DNA. How far she had advanced was reported in two articles in *Nature*[3,4] by Sir Aaron Klug, Franklin's closest collaborator at Birkbeck College, London, where she moved to from King's. He concluded that she had come very close to discovering the structure of DNA herself.

An irony of the story is that her own manuscript (coauthored by her student, R. G. Gosling and dated 17 March 1953) summarizing her results was already prepared by the time news reached King's that Watson and Crick had cracked the DNA secret. Thus she inserted a hand-written amendment to her manuscript — which was published in *Nature* on 25 April 1953 (ref. 5), along with the now-celebrated Watson and Crick paper and another by Wilkins, Herbert Wilson and Alec Stokes of King's — to say "Thus our general ideas are not inconsistent with the model proposed by Watson and Crick in the preceding communication". And so they should have been, for the Watson-Crick findings were based on her data.

There is no evidence that she knew that in late January 1953 Wilkins had innocently shown her Photograph 51, with its stark cross of black reflections (Fig. 1), to Watson, who was visiting King's. Nor did she know that in February 1953 Max Perutz, then at the Cavendish

Laboratory, had let Watson and Crick see his copy of the Medical Research Council's report summarizing the work of all principal researchers, including Franklin's.

At the same time there is no evidence that Franklin felt bitter about their achievement or had any sense of having been outrun in a race that nobody but Watson and Crick knew was a race. Indeed, she could accept the Watson–Crick model as a hypothesis only. She wrote in *Acta Crystallographica* in September 1953 that "discrepancies prevent us from accepting it in detail"[6].

Belated credit

Watson and Crick seem never to have told Franklin directly what they subsequently have said from public platforms long after her death — that they could not have discovered the double helix of DNA in the early months of 1953 without her work. This is all the more surprising in view of the close friendship that developed among the three of them — Watson, Crick and Franklin — during the remaining years of her life. During this time, she was far happier at non-sectarian Birkbeck than she ever was at King's, and led a spirited team of researchers studying tobacco mosaic virus (TMV).

From 1954 until months before her death in April 1958, she, Watson and Crick corresponded, exchanged comments on each other's work on TMV, and had much friendly contact. At Wood's Hole, Massachusetts, in the summer of 1954 Watson offered Franklin a lift across the United States as he was driving to her destination, the California Institute of Technology. In the spring of 1956 she toured in Spain with Crick and his wife Odile and subsequently stayed with them in Cambridge when recuperating from her treatments for ovarian cancer. Characteristically, she was reticent about the nature of her illness. Crick told a friend who asked that he thought it was "something female".

In the years after leaving King's, Franklin published 17 papers, mainly on the structure of TMV (including four in *Nature*). She died proud of her world reputation in the research of coals, carbons and viruses. Given her determination to avoid fanciful speculation, she would never have imagined that she would be remembered as the unsung heroine of DNA. Nor could she have envisaged that King's College London, where she spent the unhappiest two years of her professional career, would dedicate a building — the Franklin–Wilkins building — in honour of her and the colleague with whom she had been barely on speaking terms. ☐

doi:10.1038/nature01399

1. Watson, J. *The Double Helix: A Personal Account of the Discovery of the Structure of DNA* (Atheneum, New York, 1968).
2. Sayre, A. *Rosalind Franklin and DNA* (W. W. Norton & Co., New York, 1975).
3. Klug, A. Rosalind Franklin and the discovery of the structure of DNA. *Nature* **219,** 808–810, 843–844 (1968).
4. Klug, A. Rosalind Franklin and the double helix. *Nature* **248,** 787–788 (1974).
5. Franklin, R. E. & Gosling, R. G. Molecular configuration in sodium thymonucleate. *Nature* **171,** 740–741 (1953).
6. Franklin, R. E. & Gosling, R. G. The structure of sodium thymonucleate fibres: I. The influence of water content. II. The cylindrically symmetrical Patterson function. *Acta Crystallogr.* **6,** 673–677, 678–685 (1953).

Original reference: *Nature* **421,** 407–408 (2003).

The mosaic that is our genome

Svante Pääbo

*Max Planck Institute for Evolutionary Anthropology, Deutscher Platz 6, D-04103 Leipzig, Germany
(e-mail: paabo@eva.mpg.de)*

The discovery of the basis of genetic variation has opened inroads to understanding our history as a species. It has revealed the remarkable genetic similarity we share with other individuals as well as with our closest primate relatives. To understand what make us unique, both as individuals and as a species, we need to consider the genome as a mosaic of discrete segments, each with its own unique history and relatedness to different contemporary and ancestral individuals.

The discovery of the structure of DNA[1], and the realization that the chemical basis of mutations is changes in the nucleotide sequence of the DNA, meant that the history of a piece of DNA could be traced by studying variation in its nucleotide sequence found in different individuals and in different species. But it was not until rapid and inexpensive methods became available for probing DNA sequence variation in many individuals that the efficient study of molecular evolution in general — and of human evolution in particular — became feasible. Thus, the development in the 1980s of techniques for efficiently scoring polymorphisms with restriction enzymes and amplifying DNA[2,3] enabled the study of molecular evolution to become a truly booming enterprise.

What follows is a personal and, by necessity, selective attempt to consider what the accelerating pace of exploration of human genetic variation over the past two decades has taught us about ourselves as a species, as well as some suggestions for what may be fruitful areas for future studies.

Primate relations

The first insight of fundamental importance for our understanding of our origins came from comparisons of DNA sequences between humans and the great apes. These analyses showed that the African apes, especially the chimpanzees and the bonobos, but also the gorillas, are more closely related to humans than are the orangutans in Asia[4]. Thus, from a genetic standpoint, humans are essentially African apes (Fig. 1). Although there had been hints of this from molecular comparisons of proteins[5,6], it was a marked shift from the earlier common belief that humans represented their own branch separate from the great apes.

Our sense of uniqueness as a species was further rocked by the revelation that human DNA sequences differ by, on average, only 1.2 per cent from those of the chimpanzees[7], as a consequence of humans and apes sharing a recent common ancestry. It should be noted that the dating of molecular divergences has uncertainties of unknown magnitude attached, not least because of calibration based on palaeontological data. Nevertheless, it seems clear that the human evolutionary lineage diverged from that of chimpanzees about 4–6 million years ago, from that of gorillas about 6–8 million years ago, and from that of the orangutans about 12–16 million years ago[7]. Before the advent of molecular data, the human–chimpanzee divergence was widely believed to be about 30 million years old.

In fact, we have recently come to realize that the relationship between humans and the African apes is so close as to be entangled. Although the majority of regions in our genome are most closely related to chimpanzees and bonobos, a non-trivial fraction is more closely related to gorillas[7]. In yet other regions, the apes are more closely related to each other than to us (Fig. 2). This is because the speciation events that separated these lineages occurred so closely in time that genetic variation in the first ancestral species, from which the gorilla lineage diverged, survived until the second speciation event between the human and chimpanzee lineages[8]. Thus, there is not one history with which we can describe the relationship of our genome to the genomes of the African apes, but instead different histories for different segments of our genome. In this respect, our genome is a mosaic, where each segment has its own relationship to that of the African apes.

Modern humans

The mosaic nature of our genome is even more striking when we consider differences in DNA sequence between currently living humans. Our genome sequences are about 99.9 per cent identical to each other. The variation found along a chromosome is structured in 'blocks' where the nucleotide substitutions are associated in so-called haplotypes (Figs 2b and 3). These 'haplotype blocks' are likely to result from the fact that recombination, that is, the re-shuffling of chromosome segments that occurs during formation of sex cells (meiosis), tends to occur in certain areas of the chromosomes more often than in others[9–11]. In addition, the chance occurrence of recombination events at certain spots and not at others in the genealogy of human chromosomes will influence the structure of these blocks. Thus, any single human chromosome is a mosaic of different haplotype blocks, where each block has its own pattern of variation. Although the delineation of such blocks depends on the methods used to define them, they are typically 5,000–200,000 base pairs in length, and as few as four to five common haplotypes account for most of the variation in each block (Fig. 3).

Of 928 such haplotype blocks recently studied in humans from Africa, Asia and Europe[12], 51 per cent were found on all three

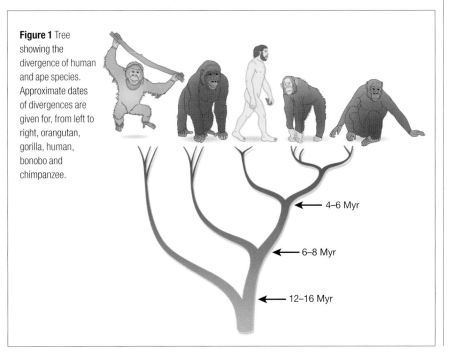

Figure 1 Tree showing the divergence of human and ape species. Approximate dates of divergences are given for, from left to right, orangutan, gorilla, human, bonobo and chimpanzee.

4–6 Myr

6–8 Myr

12–16 Myr

continents, 72 per cent in two continents and only 28 per cent on one continent. Of those haplotypes that were on one continent only, 90 per cent were found in Africa, and African DNA sequences differ on average more among themselves than they differ from Asian or European DNA sequences[13]. Therefore, within the human gene pool, most variation is found in Africa and what is seen outside Africa is a subset of the variation found within Africa.

Two parts of the human genome can be regarded as haplotype blocks where the history is particularly straightforward to reconstruct, as no recombination occurs at all. The first of these is the genome of the mitochondrion (the cellular organelle that produces energy and has its own genetic material), which is passed on to the next generation from the mother's side; the second is the Y chromosome, which is passed on from the father's side. Variation in DNA sequences from both the mitochondrial genome[14–16] and the Y chromosome[17], as well as many sections of the nuclear genome[13,18–20], have their geographical origin in Africa. Because other evidence suggest that humans expanded some 50,000 to 200,000 years ago[21] from a population of about 10,000 individuals, this suggests that we expanded from a rather small African population. Thus, from a genomic perspective, we are all Africans, either living in Africa or in quite recent exile outside Africa.

Ancient humans

What happened to the other hominids that existed in the Old World from about 2 million years ago until about 30,000 years ago? For instance, the Neanderthals are abundant in the fossil record and persisted in western Europe until less than 30,000 years ago.

Analysis of Neanderthal mitochondrial DNA has shown that, at least with respect to the mitochondrial genome, there is no evidence that Neanderthals contributed to the gene pool of current humans[22–25]. It is possible, however, that some as yet undetected interbreeding took place between modern humans and archaic hominids, such as *Homo erectus* in Asia or Neanderthals in Europe[22,26,27].

But any interbreeding would not have significantly changed our genome, as we know that the variation found in many haplotype blocks in the nuclear genome of contemporary humans is older than the divergence between Neanderthals and humans. Thus, the divergence of modern humans and Neanderthals was so recent that Neanderthal nuclear DNA sequences were probably more closely related to some current human DNA sequences than to other Neanderthals. In other words, the overlapping genetic variation that is likely to have existed between different ancient hominid forms makes it difficult to resolve the extent to which any interbreeding occurred.

Nevertheless, the limited variation among humans outside Africa, as well palaeontological evidence[28], suggest that any contribution cannot have been particularly extensive. Thus, it seems most likely that modern humans replaced archaic humans without extensive interbreeding and that the past 30,000 years of human history are unique in that we lack the company of the closely related yet distinct hominids with which we used to share the planet.

Human variation and 'race'

Comparisons of the within-species variation among humans and among the great apes have shown that humans have less genetic variation than the great apes[29,30]. Furthermore, early data that only about 10 per cent of the genetic variation in humans exist between so-called 'races'[31] is borne out by DNA sequences which show that races are not characterized by fixed genetic differences. Rather, for any given haplotype block in the genome, a person from, for example, Europe is often more closely related to a person from Africa or from Asia than to another person from Europe that shares his or her complexion (for example, see ref. 32; Fig. 2).

Claims about fixed genetic differences between races (see ref. 33 for example) have proved to be due to insufficient sampling[34]. Furthermore, because the main pattern of genetic variation across the globe is one of gene-frequency gradients[35], the contention that significant differences between races can be seen in frequencies of various genetic markers[36] is very likely due to sampling of populations separated by vast geographical distances. In this context it is worth noting that the colonization history of the United States has resulted in a sampling of the human population made up largely of people from western Europe, western Africa and southeast Asia. Thus, the fact that 'racial groups' in the United States differ in gene frequencies cannot be taken as evidence that such differences represent any true subdivision of the human gene pool on a worldwide scale.

Rather than thinking about 'populations', 'ethnicities' or 'races', a more constructive way to think about human genetic variation is to consider the genome of any particular individual as a mosaic of haplotype blocks. A rough calculation (Fig. 3) reveals that each individual carries in the order of 30 per cent of the entire haplotype variation of the human gene pool. Although not all of our

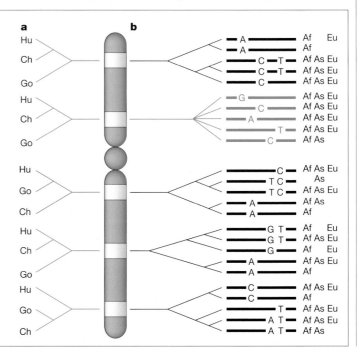

Figure 2 Within- and between-species variation along a single chromosome. **a**, The interspecies relationships of five chromosome regions to corresponding DNA sequences in a chimpanzee and a gorilla. Most regions show humans to be most closely related to chimpanzees (red) whereas a few regions show other relationships (green and blue). **b**, The among-human relationships of the same regions are illustrated schematically for five individual chromosomes. Most DNA variants are found in people from all three continents, namely Africa (Af), Asia (As) and Europe (Eu). But a few variants are found on only one continent, most of which are in Africa. Note that each human chromosome is a mosaic of different relationships. For example, a chromosome carried by a person of European descent may be most closely related to a chromosome from Asia in one of its regions, to a chromosome from Africa in another region, and to a chromosome from Europe in a third region. For one region (red), the extent of sequence variation within humans is low relative to what is observed between species. The relationship of this sequence among humans is illustrated as star-shaped owing to a high frequency of nucleotide variations that are unique to single chromosomes. Such regions may contain genes that contribute to traits that set humans apart from the apes.

Figure 3 The mosaic structure of human genetic variation. **a**, Each human chromosome is made up of regions, called 'haplotype blocks', which are stretches of DNA sequence where three to seven variants (at frequencies above 5 per cent in the human population) account for most of the variation found among humans. Each such haplotype found in a block is illustrated here as a bar of different colour. The catalogue of haplotypes for every block makes up the 'haplotype map' of the human genome. **b**, The chromosomes of two hypothetical individuals are shown. Each individual carries two copies of each block (as humans carry two sets of chromosomes). As the chance that the two haplotypes carried at a block are identical is about 20 per cent, each of us carries an average of about 1.8 different haplotypes per block. Since there is on average 5.5 haplotypes for every block, each individual carries about 30 per cent of the total haplotype diversity of the entire human species. Haplotype blocks tend to be shorter in Africa than elsewhere; as a result, African variation will probably have to be used to define the species-wide block lengths, which may be an average of around 10,000 base pairs. Note that not all of the human genome may have a clearly definable haplotype-block structure.

genome may show a typical haplotype-block structure and more research is needed to fully understand the haplotype landscape of our genome, this perspective clearly indicates that each of us contain a vast proportion of the genetic variation found in our species. In the future, we therefore need to focus on individuals rather than populations when exploring genetic variation in our species.

Tracking human traits

What are the frontiers ahead of us in human evolutionary studies? One of them, to my mind, is to identify gene variants that have been selected and fixed in all humans during the past few hundred thousand years. These will include genes involved in phenotypic traits that set humans apart from the apes and at least some archaic human forms (for example, genes involved in complex cognitive abilities, language and longevity). However, an important obstacle in this respect is that there is little detailed knowledge of many of the relevant traits in the great apes. For example, only recently has the extent to which apes possess the capability for language[37] and culture[38] begun to be comprehensively described. As a consequence, we have come to realize that almost all features that set humans apart from apes may turn out to be differences in grade rather than absolute differences.

Many such differences are likely to be quantitative traits rather than single-gene traits. To have a chance to unravel the genetic basis of such traits, we will need to rigorously define the differences between apes and humans — for instance, how we learn, how we communicate and how we age. In the next few years, geneticists will therefore need to consider insights from primatology and psychology, and more studies will be required that directly compare humans to apes.

There are, however, ways in which we can contribute towards the future unravelling of functionally important genetic differences between humans and apes. For example, we can identify regions of the human genome where the patterns of variation suggest the recent occurrence of a mutation that was positively selected and swept through the entire human population. The sequencing of the chimpanzee genome, as well as the haplotype-map project, will greatly help in this. Further prerequisites include the capability to determine the DNA sequence of many human genomes and the development of tools and methods to analyse the resulting data; in particular, a more realistic model of human demographic history is required.

Collectively these studies will allow us to identify regions in the human genome that have recently been acted upon by selection and thus are likely to contain genes contributing to human-specific traits (Fig. 2). Other interesting candidate genes for human-specific traits are genes duplicated or deleted in humans[39], genes that have changed their expression in humans[40], and genes responsible for disorders affecting traits unique to humans, such as language[41] and a large brain size[42].

A problem inherent in studying genes that are involved in traits unique to humans, such as language, is that functional experiments cannot be performed, as no animal model exists, and transgenic humans or chimpanzees cannot be constructed. A further difficulty is that many genes that enable humans to perform tasks of interest may exert their effects during early development where our ability to study their expression both in apes and humans is extremely limited.

A challenge for the future is therefore to design ways around these difficulties. This will involve *in vitro* as well as *in silico* approaches that study how genes interact with each other to influence developmental and physiological systems. As these goals are achieved, we will be able to determine the order and approximate times of genetic changes during the emergence of modern humans that led to the traits that set us apart among animals. □

doi:10.1038/nature01400

1. Watson, J. D. & Crick, F. H. C. A structure for deoxyribose nucleic acid. *Nature* **171**, 737–738 (1953).
2. Botstein, D., White, R. L., Skolnick, M. & Davis, R. W. Construction of a genetic linkage map in man using restriction fragment length polymorphisms. *Am. J. Hum. Genet.* **32**, 314–331 (1980).
3. Saiki, R. K. *et al.* Enzymatic amplification of β-globin genomic sequences and restriction site analysis for diagnosis of sickle cell anemia. *Science* **230**, 1350–1354 (1985).
4. Miyamoto, M. M., Slightom, J. L. & Goodman, M. Phylogenetic relations of humans and African apes from DNA sequences in the ψη-globin region. *Science* **238**, 369–373 (1987).
5. Mayr, E. *Animal Species and Evolution* (Harvard Univ. Press, Cambridge, MA, 1963).
6. Wilson, A. C. & Sarich, V. M. A molecular time scale for human evolution. *Proc. Natl Acad. Sci. USA* **63**, 1088–1093 (1969).
7. Chen, F. C., Vallender, E. J., Wang, H., Tzeng, C. S. & Li, W. H. Genomic divergence between human and chimpanzee estimated from large-scale alignments of genomic sequences. *J. Hered.* **92**, 481–489 (2001).
8. Nei, M. *Molecular Evolutionary Genetics* (Columbia Univ. Press, New York, 1987).
9. Daly, M. J., Rioux, J. D., Schaffner, S. F., Hudson, T. J. & Lander, E. S. High-resolution haplotype structure in the human genome. *Nature Genet.* **29**, 229–232 (2001).
10. Jeffreys, A. J., Kauppi, L. & Neumann, R. Intensely punctated meiotic recombination in the class II region of the major histocompatibility complex. *Nature Genet.* **29**, 217–222 (2001).
11. Patil, N. *et al.* Blocks of limited haplotype diversity revealed by high-resolution scanning of human chromosome 21. *Science* **294**, 1719–1723 (2001).
12. Gabriel, S. B. *et al.* The structure of haplotype blocks in the human genome. *Science* **296**, 2225–2229 (2002).
13. Yu, N. *et al.* Larger genetic differences within Africans than between Africans and Eurasians. *Genetics* **161**, 269–274 (2002).
14. Cann, R. L., Stoneking, M. & Wilson, A. C. Mitochondrial DNA and human evolution. *Nature* **325**, 31–36 (1987).
15. Vigilant, L., Stoneking, M., Harpending, H., Hawkes, K. & Wilson, A. C. African populations and the evolution of human mitochondrial DNA. *Science* **253**, 1503–1507 (1991).
16. Ingman, M., Kaessmann, H., Pääbo, S. & Gyllensten, U. Mitochondrial genome variation and the origin of modern humans. *Nature* **408**, 708–713 (2000).
17. Underhill, P. A. *et al.* Y chromosome sequence variation and the history of human populations. *Nature Genet.* **26**, 358–361 (2000).
18. Stoneking, M. *et al.* Alu insertion polymorphisms and human evolution: evidence for a larger population size in Africa. *Genome Res.* **7**, 1061–1071 (1997).
19. Tishkoff, S. A. *et al.* Global patterns of linkage disequilibrium at the CD4 locus and modern human origins. *Science* **271**, 1380–1387 (1996).
20. Takahata, N., Lee, S. H. & Satta, Y. Testing multiregionality of modern human origins. *Mol. Biol. Evol.* **18**, 172–183 (2001).
21. Harpending, H. & Rogers, A. Genetic perspectives on human origins and differentiation. *Annu. Rev. Genomics Hum. Genet.* **1**, 361–385 (2000).
22. Krings, M. *et al.* Neanderthal DNA sequences and the origin of modern humans. *Cell* **90**, 19–30 (1997).
23. Ovchinnikov, I. V. *et al.* Molecular analysis of Neanderthal DNA from the northern Caucasus. *Nature* **404**, 490–493 (2000).
24. Krings, M., Geisert, H., Schmitz, R. W., Krainitzki, H. & Pääbo, S. DNA sequence of the mitochondrial hypervariable region II from the Neandertal type specimen. *Proc. Natl Acad. Sci. USA* **96**, 5581–5585 (1999).

25. Krings, M. *et al.* A view of Neandertal genetic diversity. *Nature Genet.* **26**, 144–146 (2000).
26. Nordborg, M. On the probability of Neanderthal ancestry. *Am. J. Hum. Genet.* **63**, 1237–1240 (1998).
27. Pääbo, S. Human evolution. *Trends Cell Biol.* **9**, M13–M16 (1999).
28. Stringer, C. Modern human origins: progress and prospects. *Phil. Trans. R. Soc. Lond. B* **357**, 563–579 (2002).
29. Deinard, A. & Kidd, K. Evolution of a HOXB6 intergenic region within the great apes and humans. *J. Hum. Evol.* **36**, 687–703 (1999).
30. Kaessmann, H., Wiebe, V., Weiss, G. & Pääbo, S. Great ape DNA sequences reveal a reduced diversity and an expansion in humans. *Nature Genet.* **27**, 155–156 (2001).
31. Lewontin, R. C. The problem of genetic diversity. *Evol. Biol.* **6**, 381–398 (1972).
32. Kaessmann, H., Heissig, F., von Haesler, A. & Pääbo, S. DNA sequence variation in a non-coding region of low recombination on the human X chromosome. *Nature Genet.* **22**, 78–81 (1999).
33. Harris, E. E. & Hey, J. X chromosome evidence for ancient human histories. *Proc. Natl Acad. Sci. USA* **96**, 3320–3324 (1999).
34. Yua, N. & Li, W.-H. No fixed nucleotide difference between Africans and non-Africans at the pyruvate dehydrogenase E1 α-subunit locus. *Genetics* **155**, 1481–1483 (2000).
35. Cavalli-Sforza, L. L., Menozzi, P. & Piazza, A. *The History and Geography of Human Genes* (Princeton Univ. Press, Princeton, NJ, 1993).
36. Risch, N., Burchard, E., Ziv, E. & Tang, H. Categorization of humans in biological research: genes, race and disease. *Genome Biol.* **3**, 2007.1–2007.12 (2002).
37. Tomasello, M. & Call, J. *Primate Cognition* (Oxford Univ. Press, New York, 1997).
38. Whiten, A. *et al.* Cultures in chimpanzees. *Nature* **399**, 682–685 (1999).
39. Eichler, E. E. Recent duplication, domain accretion and the dynamic mutation of the human genome. *Trends Genet.* **17**, 661–669 (2001).
40. Enard, W. *et al.* Intra- and interspecific variation in primate gene expression patterns. *Science* **296**, 340–343 (2002).
41. Enard, W. *et al.* Molecular evolution of *FOXP2*, a gene involved in speech and language. *Nature* **418**, 869–872 (2002).
42. Jackson, A. P. *et al.* Identification of microcephalin, a protein implicated in determining the size of the human brain. *Am. J. Hum. Genet.* **71**, 136–142 (2002).

Acknowledgements
My work is funded by the Max Planck Society, the Bundesministerium für Bildung und Forschung and the Deutsche Forschungsgemeinschaft. I thank B. Cohen, H. Kaessmann, D. Serre, M. Stoneking, C. Stringer, L. Vigilant and especially D. Altshuler for helpful comments on the manuscript.

Original reference: *Nature* **421**, 409–412 (2003).

Nature, nurture and human disease

Aravinda Chakravarti* & Peter Little†

**McKusick-Nathans Institute of Genetic Medicine, Johns Hopkins University School of Medicine, 600 North Wolfe Street, Jefferson Street Building, 2-109, Baltimore, Maryland 21287, USA (e-mail: aravinda@jhmi.edu)*
†*School of Biotechnology and Biomolecular Sciences, University of New South Wales, Sydney, New South Wales 2052, Australia (e-mail: p.little@unsw.edu.au)*

What has been learnt about individual human biology and common diseases 50 years on from the discovery of the structure of DNA? Unfortunately the double helix has not, so far, revealed as much as one would have hoped. The primary reason is an inability to determine how nurture fits into the DNA paradigm. We argue here that the environment exerts its influence at the DNA level and so will need to be understood before the underlying causal factors of common human diseases can be fully recognized.

"We used to think our fate was in our stars. Now we know, in large measure, our fate is in our genes." J. D. Watson, quoted in *Time* magazine, 20 March 1989 (ref. 1).

The double helix, in its simplicity and beauty, is the ultimate modern icon of contemporary biology and society. Its discovery provided the bridge between the classical breeding definition and the modern functional definition of genetics, and permanently united genetics with biochemistry, cell biology and physiology. The DNA structure provided an immediate explanation for mutation and variation, change, species diversity, evolution and inheritance. It did not, however, automatically provide a mechanism for understanding how the environment interacts at the genetic level.

One gene, one disease
Recognition that genes have a role in human disease dates back to the rediscovery of the rules that govern the inheritance of genes by Gregor Mendel — the so-called Mendelian laws of inheritance. So far, human geneticists have been most successful at understanding single-gene disorders, as their biological basis, and thus presumed action, could be predicted from inheritance patterns. Mendelian diseases are typically caused by mutation of a single gene that results in an identifiable disease state, the inheritance of which can readily be traced through generations.

The landmark sequencing of the human genome provided some important lessons about the role of genes in human disease. Notably, mutations in specific genes lead to specific biological changes, and rarely do mutations in multiple genes lead to an identical set of characteristics that obey 'Mendelian inheritance'. Additionally, sequence diversity of mutations is large and, consequently, individual mutations are almost always rare, showing relatively uniform global distributions.

But a few exceptions do exist. Some recessive mutations (mutations that influence a person only if both copies of the gene are altered) are surprisingly common in specific populations. This defiance of general mutation patterns arises either from chance increases in frequency in isolated populations, such as in the Old Order Amish[2], or from the protective effect of a deleterious mutation in a single copy, such as the genetic mutation that on the one hand causes sickle-cell anaemia, but on the other hand offers protection against malaria[3]. These examples show that human history, geography and ecology of a particular people are relevant to understanding their present-day molecular disease burden[4].

For over 90 years, the association between DNA mutations and a vast variety of single-gene disorders has repeatedly emphasized the notion that human disease results from faults in the DNA double helix (see, for example, the Online Mendelian Inheritance in Man database at www.ncbi.nlm.nih.gov/omim/, which provides a catalogue of human genes and genetic disorders). Is it then too extrapolative to suggest that all diseases and traits, each of which has some familial and imputed inherited component, will be caused by a corrupted piece of double helix?

Is our fate encoded in our DNA?
Is Watson's genetic aphorism of human disease really true? The excitement of genetics, and the perceived medical importance of the human genome sequence, is pegged to the promise of an understanding of common chronic disease and not rare Mendelian diseases. In theory, one might hope that approaches used successfully to identify single-gene diseases could simply be applied to the common causes of world-wide morbidity and mortality, such as cancer, heart disease, psychiatric illness and the like. This would enable a boon for diagnosis, understanding and the eventual treatment of these common maladies[5].

The reality is that progress towards identifying common disease mutations has been slow, and only recently have there been some successes[6]. It is now appreciated that although genes are one contributor to the origin of common diseases, the mutations they contain must have properties that are different from the more familiar, deterministic features of single-gene mutations. Indeed, the underlying genes are likely to be numerous, with no single gene having a major role, and mutations within these genes being common and imparting small genetic effects (none of which are either necessary or sufficient[7]).

Moreover, there is a suspicion that these mutations both interact with one another and with the environment and lifestyle, although the molecular specificity of interactions is unproven[8]. To complicate matters, common disorders frequently show large population differences that have led to health disparities and, as is becoming more evident, the incidence of these disorders can show significant changes over time[9].

Interplay of DNA and environment

The inability of geneticists to easily identify common disease genes has been seen as a vindication of the importance of nurture. This is too simplistic; the influence of nature and nurture cannot be neatly divided, as it is clear that nurture is important to biology through its actions on DNA and its products. The environment must affect the regulation of critical genes by some mechanism and so, seen another way, mutations are not the only agent for altering gene function.

The scientific literature of cancer research reveals that despite having heterogeneous origins — both inherited and acquired — a specific tumour develops only from altering the expression (activity) of specific sets of genes[10]. That is, a variety of exposures and mutations collaborate to change the activity of specific genes and, consequently, interrupt precise aspects of cell metabolism. The regulation of circadian rhythm is another example of how external environmental cues influence DNA functions[11].

Thus, the double helix inevitably interacts with the environment, directly and indirectly, to predispose or protect us from disease. If perturbations of multiple genes contribute to a disorder, then the activities of these genes can be affected by any combination of mutation and environmental exposure altering their function. It is our opinion that genes have a stronger, maybe even a pervasive, role in all diseases and traits, with the understanding that it is the collective action of genes and nurture that underpins ultimate disease outcome.

Rather than dismissing the role of environment, our view embraces it directly, and, by that, expands the meaning of the term 'genetic'. It also emphasizes the work that remains to be done to understand gene regulation and, in particular, how genes and their products are modulated by external cues and how homeostasis is disrupted in human disease. Human beings are each the product of a unique genome and a unique set of experiences. Both need to be understood to intervene effectively in disease causation.

Implications for medicine

What does this mean in practice? The assessment of the quantitative role of genes in human traits is derived largely from studies on identical and fraternal twins (Fig. 1). By this measure, all common disorders have a 'genetic' basis, but the contribution varies from slight in some cancers and multiple sclerosis,

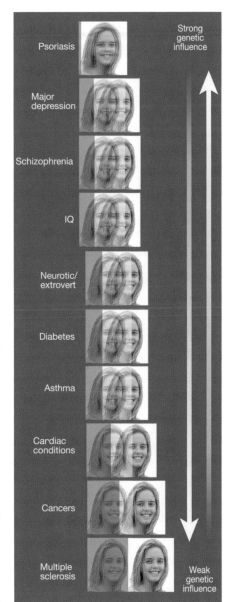

Figure 1 Studies of identical twins have revealed that some conditions, such as psoriasis, have a strong genetic component and are less influenced by environmental and lifestyle factors — identical twins are more likely to share these diseases. But other conditions, such as multiple sclerosis, are only weakly influenced by genetic makeup and therefore twins may show differences depending on their exposure to various environmental factors.

to moderate in diabetes, heart diseases, migraine and asthma, to high in disorders such as psoriasis[12]. Critically, the discordance between identical twins — where twins show different diseases despite being genetically identical — illustrates the influence of exogenous factors, but does not prove the lack of influence of genes: of course, environmental factors over a lifetime affect an individual's chance of developing disease.

Let us assume, for the sake of argument, that all of the relevant genetic and environmental factors are identified that lead to a disease. Appreciating the relationship of

genetic variation and environment suggests that a number of presently fashionable ideas about genetics are simplistic; two in particular are the 'bar code' view of genetic diagnosis and the 'right medicine for the right patients'.

Common genetic variations are essentially binary — either an adenine or guanine base, or a cytosine or thymine base — at a given position in the sequence. Unfortunately, this leads to a tendency to define genetic individuality as a binary pattern, a so-called 'bar code' for each individual. Some genetic variants convey susceptibility to a disease, but they typically convey risk rather than certainty of being afflicted with a condition.

Knowledge based on the sequence could have significant public health implications, and even be predictive at the population level. But a human DNA bar code would provide uncomfortable, perhaps even intolerable, knowledge of likely outcomes, with no certainty, only probabilities. Most individuals, we suspect, are ill equipped to deal with the knowledge that they have a 50 per cent chance of succumbing to an illness; equally, society has had great difficulty in knowing how to respond to such information, hence the concerns regarding genetic discrimination[13]. The reality is that the genetic bar code is weakly predictive and individuals may find this threatening, life enhancing or just irrelevant; in any event, much work is needed to enable the predictive revolution in medicine.

Human genetic individuality has forced the recognition that medicine has to refocus on the individual. This has been the rallying cry, particularly within the pharmaceutical business, of pharmacogenomics (the application of genome-scale understanding to the development of medicines), and there is no doubt that understanding of the variation within drug-metabolizing enzymes has exploded in the past 20 years[14]. The underpinning idea is enormously attractive — if genetic analysis of key DNA variations can be used to understand how individuals might respond to drugs, then it could be possible to eliminate the difficult, sometimes lethal, hit-and-miss approaches to medication that are a necessary feature of present medical practice.

Unfortunately, the influence of lifestyle is just as much a feature of drug response as it is of any other genetically influenced condition. The classic case of the influence of drinking grapefruit juice on the levels of many drugs[15] illustrated that there can be no such thing as 'the patient', because the patient is living in a complex world that changes by the minute. Once again, predictions for the population do not have the same predictive power for individuals.

Future challenges

The challenges that lifestyle presents to genetic studies are considerable. We believe that the next 50 years will bring a genuine revolution of far greater individual significance than that delivered by genetics over the past 50 years. This is because lifestyle can conceivably be analysed, and in so doing, it should be possible to develop a genuinely personalized medicine.

Researchers can now think seriously about how to identify lifestyle influences: such studies will have to be on an unprecedented scale and one of the first of these, proposed to comprise 500,000 individuals in the United Kingdom, has already started[16]. These kinds of studies are a bold venture into relatively uncharted territory and face substantial technical, biological and science-culture challenges.

Scientifically, it is necessary to understand a deceptively simple equation: genes + environment = outcome. The difficulty here is the uncertainty surrounding both terms in the equation; ideally, one set of genetic factors will interact with one set of environmental influences to produce identical outcomes, but it is unknown whether this is always going to be the case. A far more difficult relationship would exist if multiple genetic factors interacted with multiple environments to achieve the same outcome. The example of glutathione S-transferase mutations, smoking and incidence of lung cancer[17] shows it is possible to detect some interactions, but it is unclear how, or even if, statistical methods might be developed for addressing the more complex possibilities.

Perhaps the greatest unknown in undertaking these projects is human psychology; the consequences of smoking have been known for many decades, but people still smoke. Advice does not imply acceptance. How to turn knowledge into practical outcomes must be an increasing focus of attention for both researchers and funding agencies.

Psychology is also in play in the initial decision to undertake this research; for researchers, funding agencies and politicians there is great risk implicit in undertaking a hugely expensive project with complex outcome. People would like to live in a simpler world, with simpler decisions, but the vision of such a project is enormous: once complete, as much will be known about the origins of human disorders as can be discovered by using such epidemiological and genetic studies. Perhaps more important, the beginnings of a new medicine will emerge, one focused uniquely and completely upon the individual, upon the combination of genetic uniqueness and personal choices that are the very essence of individual lives.

If we are collectively bold in our present decisions and accept the risk of action, a world can be created where medicine is a guide, not a place of last resort. If the past 50 years has seen the revolution of DNA, then the revolution cannot be completed without an appreciation of both genetic and environmental individuality; only then will individuals understand the meaning of their inheritance. ☐

doi:10.1038/nature01401

1. Jaroff, L. The gene hunt. *Time* 20 March, 62–67 (1989)
2. Arcos-Burgos, M. & Muenke, M. Genetics of population isolates. *Clin. Genet.* 61, 233–247 (2002).
3. Aidoo, M. *et al.* Protective effects of the sickle cell gene against malaria morbidity and mortality. *Lancet* 359, 1311–1312 (2002).
4. Tishkoff, S. A. & Williams, S. M. Genetic analysis of African populations: human evolution and complex disease. *Nature Rev. Genet.* 3, 611–621 (2002).
5. Chakravarti, A. Single nucleotide polymorphisms: ...to a future of genetic medicine. *Nature* 409, 822–823 (2001).
6. Carrasquillo, M. M. *et al.* Genome-wide association study and mouse model identify interaction between RET and EDNRB pathways in Hirschsprung disease. *Nature Genet.* 32, 237–244 (2002).
7. Cox, N. J. Challenges in identifying genetic variation affecting susceptibility to type 2 diabetes: examples from studies of the calpain-10 gene. *Hum. Mol. Genet.* 10, 2301–2305 (2001).
8. Sullivan, P. F. *et al.* Analysis of epistasis in linked regions in the Irish study of high-density schizophrenia families. *Am. J. Med.*
Genet. 105, 266–270 (2001).
9. Boyle, J. P. *et al.* Projection of diabetes burden through 2050: impact of changing demography and disease prevalence in the United States. *Diabetes Care* 24, 1936–1940 (2001).
10. Münger, K. Disruption of oncogene/tumor suppressor networks during human carcinogenesis. *Cancer Invest.* 20, 71–81 (2002).
11. Panda, S., Hogenesch, J. B. & Kay, S. A. Circadian rhythms from flies to human. *Nature* 417, 329–335 (2002).
12. MacGregor, A. J. *et al.* Twins: novel uses to study complex traits and genetic diseases. *Trends Genet.* 16, 131–134 (2000).
13. Wertz, D. C. Ethics watch. *Nature Rev. Genet.* 3, 496 (2002).
14. Evans, W. E. & Johnson, J. A. Pharmacogenomics: the inherited basis for interindividual differences in drug response. *Annu. Rev. Genomics Hum. Genet.* 2, 9–39 (2001).
15. Lown, K. S. *et al.* Grapefruit juice increases felodipine oral availability in humans by decreasing intestinal CYP3A protein expression. *J. Clin. Invest.* 99, 2545–2553 (1997).
16. Wright, A. F., Carothers, A. D. & Campbell, H. Gene-environment interactions—the BioBank UK study. *Pharmacogenomics J.* 2, 75–82 (2002).
17. Stucker, I. *et al.* Genetic polymorphisms of glutathione S-transferases as modulators of lung cancer susceptibility. *Carcinogenesis* 23, 1475–1481 (2002).

Original reference: *Nature* 421, 412–414 (2003).

The double helix in clinical practice

John I. Bell

The Office of the Regius Professor of Medicine, University of Oxford, Oxford OX3 9DU, UK
(e-mail: regius@medsci.ox.ac.uk)

The discovery of the double helix half a century ago has so far been slow to affect medical practice, but significant transformations are likely over the next 50 years. Changes to the way medicine is practised and new doctors are trained will be required before potential benefits are realized.

"It is much more important to know what kind of patient has a disease than to know what kind of disease a patient has." Caleb Parry, 18th century physician, Bath.

The structure of DNA established the basic framework that would develop into the field of molecular genetics. The information gleaned from this scientific endeavour continues to have a profound influence on our understanding of biological systems[1]. As most human diseases have a significant heritable component, it was soon recognized that the characterization of the genetic determinants of disease would provide remarkable opportunities for clinical medicine, potentially altering the way disease was understood, diagnosed and treated.

But despite the obvious potential applications to medicine, the development of significant genetic advances relevant to clinical practice could take generations. This is in marked contrast to many other medically related discoveries that occurred around the same time and which were translated rapidly into clinical practice. For instance, the development of penicillin by Ernst Chain and Howard Florey in 1941 was saving thousands of lives within months of their discovery of how to efficiently produce the antibiotic[2]. Discoveries relating to disease aetiology, such as the recognition in 1950 of a relationship between smoking and lung cancer, have had a profound effect on mortality[3]. This was despite the convictions of at least one distinguished statistical geneticist who argued against the causality of this observation, implying that a common genetic factor caused both lung cancer and a predilection to smoking cigarettes[4]!

Although other important discoveries have had demonstrably more impact on health care at the time of their fiftieth anniversaries than has the double helix, its slower transition from discovery to clinical implementation will be balanced by its potentially profound impact across all medical disciplines. Progress has been slow, but mounting evidence suggests that, while public health and antibiotics produced important healthcare outcomes in the past 50 years, the next 50 are likely to belong to genetics and molecular medicine.

The potential impact of genetics on clinical practice has been questioned by some observers[5] who believe that the positive predictive value of genetic testing for most common disease genes will be insufficient to provide the beneficial effects seen with single-gene disorders, which affect only a tiny proportion of the population. Many advocates of genetics argue, on the other hand, that our understanding of disease is

The caduceus — Hermes' winged staff entwined with two snakes — represents a symbol long adopted by medicine.

undergoing a major change. They contend that genetic research is playing a fundamental role in improving our understanding of the pathophysiology that underlies disease and that, inevitably, as this is applied, it will alter both the theory and practice of medicine in the future[6].

A new taxonomy for human disease

Clinical practice has always been limited by its inability to differentiate clinical, biochemical and pathological abnormalities that accompany a disease from those events actually responsible for mediating a disease process. Clinicians may have moved on from calling 'fever' a disease[7], but they still rely on phenotypic criteria to define most diseases, and yet these may obscure the underlying mechanisms and often mask significant heterogeneity. As Thomas Lewis pointed out in 1944, diagnosis of most human disease provides only "insecure and temporary conceptions"[8]. Of the main common diseases, only the infectious diseases have a truly mechanism-based nomenclature.

An understanding of the genetic basis of maladies is providing a new taxonomy of disease, free from the risk that the diagnostic criteria related to events are secondary to the disease process, rather than to its cause. Genetic information has allowed us to identify mechanistically distinct forms of diabetes, defining an autoimmune form of the disease associated with human leukocyte antigens (a highly diverse complex of immune-system genes), and recently has implicated dysfunction of factors that affect both expression and modification of gene products in mediating the adult form of the disorder[9]. Similarly, we are now aware of a range of molecules and pathways previously not recognized in the pathogenesis of asthma[10–12].

A clearer understanding of the mechanisms and pathways that mediate disease will lead to the definition of distinct disease subtypes, and may resolve many questions relating to variable disease symptoms, progression and response to therapy seen within current diagnostic categories. Ultimately, this may provide the greatest contribution genetics will make to clinical practice: a new taxonomy for human disease.

A medical revolution

Knowing that a disease can arise from a distinct mechanism will alter a physician's approach to a patient with that disorder, allowing a more accurate prognosis and choice of the most appropriate therapy. The gene 'mutations' responsible for many single-gene disorders are now commonly used in diagnostic practice, whereas those associated with common complex diseases are just being characterized. Although their predictive value will be less than with single-gene disorders, their contribution as risk factors will be similar to other risk factors such as blood pressure, cholesterol levels and environmental exposures. Because much of clinical practice involves evaluating and acting on risk probabilities, the addition of genetic risk factors to this process will be an important extension of existing practice. The overall effect of genetic risk factors is likely to be significant. For example, recent estimates in breast cancer suggest that the attributable genetic risks are likely to exceed the predictive value of a range of existing non-genetic risk factors[13].

Other potential applications of genetics in health care may be realized in a shorter timeframe. Individual variation in response to drugs and in drug toxicity is a significant problem, both in clinical practice and in the development of new therapeutic agents. Clear examples now exist of genetic variants that alter metabolism, drug response or risk of toxicity[14,15]. Such information provides an opportunity to direct therapy at individuals most likely to benefit from an intervention, thereby reducing cost and toxicity, and improving methods for drug development.

The discovery of the structure of DNA not only led to an ability to characterize genetic determinants in disease, but also provided the tools necessary for the revolution in molecular medicine that has occurred in the past 25 years. The description of the double helix was the first important step in the development of techniques to cut, ligate and amplify DNA. The application of these molecular biology and DNA-cloning techniques has already had a profound impact on our understanding of the basic cellular and molecular processes that underlie disease.

Molecular biology has improved our ability to study proteins and pathways involved in disease and has provided the technology necessary to generate new sets of targets for small-molecule drug design. It has also enabled the creation and production of a new range of biological therapeutics — recombinant proteins such as interferon, erythropoietin and insulin, as well as therapeutic antibodies, which are one of the fastest growing classes of new treatments. Further extensions of this methodology will see the inevitable introduction of DNA-based therapies that will produce proteins of interest in the appropriate cellular setting. DNA-based vaccines represent the first wave of such novel gene-therapy approaches to disease and many more are expected to follow.

We are undergoing a revolution in clinical practice that depends upon a better understanding of disease mechanisms and pathways at a molecular level. Much has already been achieved: an enhanced understanding of disease-related pathways, new therapies, novel approaches to diagnostics and new tools for identifying those at risk. But more remains to be done before the full impact of genetics on medicine is realized. Complex disease, with multiple susceptibility determinants (both environmental and genetic), will take time to dissect. This information must then be moved into the clinic and evaluated for its benefits.

As the practice of medicine moves to one more scientifically founded in disease mechanisms, many aspects of clinical practice will need to be transformed. Individual genetic variation is likely to explain a significant part of the heterogeneity seen clinically in the natural history of disease and in response to therapy. Tools to tailor medicine to an individual's needs rather than directing it at a population will inevitably become available. Similarly, as predictions of risk improve, early or preventative therapy of high-risk populations will become a reality, with screening programmes targeted to those at particularly high risk.

Transforming clinical practice

For fundamental changes to take place in clinical practice, sweeping transformation will be needed to healthcare provision, economic management and training. It is currently difficult to predict the cost–benefit ratio for such changes — certainly the present impact of molecular medicine has not made medicine less expensive. Few medical schools adequately train their students to think mechanistically about disease; indeed, the trend towards pattern-recognition medicine, away from basic science training, means that we are still far from educating the next generation of clinicians to apply the knowledge and tools bequeathed to us by the double helix. The evolution in health care that will incorporate these new principles of early diagnosis and individualized therapy will be a daunting

Table 1 **Molecular genetics in clinical practice**
• Mechanistically based diagnostic criteria
• Predisposition testing and screening
• Rapid molecular diagnostic testing of pathogens
• Pharmacogenetics
• Identification of new drug targets
• Tools for molecular medicine (for example, recombinant DNA methodology)
• Recombinant expression of therapeutic proteins
• Gene therapy

challenge in an era of uncertainty for healthcare systems worldwide.

The influence of genetic and molecular medicine on the health of patients is already sufficiently ubiquitous that it will have an impact on most common diseases. Its influence will grow over the next few decades (Table 1). It will not, however, answer all of the questions about human health, nor will it provide all the answers for optimizing clinical practice. The reductionism that accompanies molecular genetics will identify the pieces in the jigsaw, but assembling these to understand how complex systems malfunction will require a substantially more integrated approach than is available at present.

The crucial role played by environmental determinants of disease will perhaps become more tractable when combined with an understanding of genetic susceptibility. Sceptics, rightly, will wish to see more data before they acknowledge that molecular medicine will be truly transformed over the next 50 years, despite the fact that its influence on diagnostics and new therapeutics is already clearly apparent. A transition is underway, the direction of travel is clear, but managing the change in clinical practice may prove at least as challenging as resolving the original structure of the helix. ☐

doi:10.1038/nature01402

1. Watson, J. D. & Crick, F. C. H. A structure for deoxyribose nucleic acid. *Nature* **171,** 737–738 (1953).
2. Abraham, E. P. *et al.* Further observations on penicillin. *Lancet* **ii,** 177–188 (1941).
3. Doll, R. & Hill, A. B. Smoking and carcinoma of the lung. *Br. Med. J.* **2,** 739–748 (1950).
4. Fisher, R. A. Cancer and smoking. *Nature* **182,** 596 (1958).
5. Holtzman, N. A. & Marteau, T. M. Will genetics revolutionise medicine? *N. Engl. J. Med.* **343,** 141–144 (2000).
6. Bell, J. I. The new genetics in clinical practice. *Br. Med. J.* **316,** 618–620 (1998).
7. Osler, W. *The Principles and Practice of Medicine* (Appleton, New York, 1892).
8. Lewis, T. Reflections upon medical education. *Lancet* **i,** 619–621 (1944).
9. Cardon, L. R. & Bell, J. I. Association study designs for complex diseases. *Nature Rev. Genet.* **2,** 91–99 (2001).
10. Van Eerdewegh, P. *et al.* Association of the *ADAM33* gene with asthma and bronchial hyperresponsiveness. *Nature* **418,** 426–430 (2002).
11. Cookson, W. O. C. M., Sharp, P. A., Faux, J. A. & Hopkin, J. M. Linkage between immunoglobulin E responses underlying asthma and rhinitis and chromosome 11q. *Lancet* **1,** 1292–1295 (1989).
12. Shirakawa, I. *et al.* Atopy and asthma: genetic variants of IL-4 and IL-13 signalling. *Immunol. Today* **21,** 61–64 (2000).
13. Pharoah, P. D. P. *et al.* Polygenic susceptibility to breast cancer and implications for prevention. *Nature Genet.* **31,** 33–36 (2002).
14. Splawski, I. *et al.* Variant of SCN5A sodium channel implicated in risk of cardiac arrhythmia. *Science* **297,** 1333–1336 (2002).
15. Weber, W. W. *Pharmacogenetics* (Oxford Univ. Press, New York, 1997).

Original reference: *Nature* **421,** 414–416 (2003).

The *Mona Lisa* of modern science

Martin Kemp

Department of the History of Art, University of Oxford, Littlegate House, St Ebbes, Oxford OX1 1PT, and Wallace Kemp, Artakt, Studio D, 413-419, Harrow Road, London W9 3QJ, UK
(e-mail: martin.kemp@trinity.ox.ac.uk)

No molecule in the history of science has reached the iconic status of the double helix of DNA. Its image has been imprinted on all aspects of society, from science, art, music, cinema, architecture and advertising. This review of the *Mona Lisa* of science examines the evolution of its form at the hands of both science and art.

"A monkey is a machine that preserves genes up trees, a fish is a machine that preserves genes in water; there is even a small worm that preserves genes in German beer mats. DNA works in mysterious ways." Richard Dawkins in *The Selfish Gene* (Oxford University Press, 1976).

History has thrown up a few super-images, which have so insinuated themselves into our visual consciousness that they have utterly transcended their original context. This is epitomized by the *Mona Lisa*, painted by Leonardo da Vinci around 1503. The double helix of DNA is unchallenged as the image epitomizing the biological sciences. Both images speak to audiences far beyond their respective specialist worlds, and both carry a vast baggage of associations.

In the worlds of popular image diffusion, particularly on the Internet, the double helix is beginning to rival the *Mona Lisa* as a playground for eccentrics and obsessives (Fig. 1). There is an apparent difference, of course. Leonardo's panel painting is the product of human artifice, whereas DNA is a naturally occurring, large organic molecule. But Leonardo claimed that his art represented a systematic remaking of nature on the basis of a rational understanding of causes and effects. His painting is the result of a complex, nonlinear interaction between concept, subject, plan of action, acquired knowledge, skill, medium and the evolving image itself. In *The Art of Genes*[1], Enrico Coen argues that "biological development and human creativity are highly interactive processes in which events unfold rather than being necessarily pre-planned or anticipated. In other words, in both cases there is no easy separation between plan (or programme) and execution."

Looking at the investigation and representations of the double helix, we can say that they are cultural activities no less than any painting. Behind the discovery lies the vast infrastructure of a scientific culture that led to the development of the knowledge, theories, institutions, techniques and equipment that made the quest both possible and desirable. The very natures of scientific models and representations, using whatever technique, are integral to the vehicles of science communication. Their visual look is compounded from a complex set of factors, ranging from technical to aesthetic. But, in case anyone should be getting the wrong impression, I acknowledge that the cultural vehicles are designed to deliver non-arbitrary information that is open to rational scrutiny as a way of working towards real knowledge of the physical constitution of the world.

Looked at from a popular perspective (and even from the standpoint of reputation within science), James Watson and Francis Crick are identified with DNA no less than Leonardo is identified with the *Mona Lisa*. The researchers were in a very real sense the 'authors' or 'artists' of the acts of visualization that generated their models of the molecule. But their brilliant achievement was not necessarily of a higher order than that of the other pioneers of molecular modelling, such as the Braggs, John Kendrew, Max Perutz, Maurice Wilkins and Linus Pauling. Rather, they were uniquely fortunate that their molecule was both visually compelling, as a supreme example of nature's 'sculpture', and

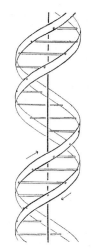

Figure 2 Structure of DNA, drawn by Francis Crick's wife Odile Crick, which was published as the sole figure in Watson and Crick's seminal paper in *Nature*, 25 April 1953 (ref. 2).

Figure 1 LEGO model of the DNA double helix (in reverse!) by Eric Harshbarger (2001), who also used his mastery of the coloured units of LEGO to compose a 'pixelated' LEGO version of the *Mona Lisa*. (Images courtesy of E. Harshbarger.)

lay at the heart of the twentieth-century version of the quest to unravel the ultimate secret of life.

The 50-year journey of the DNA molecule from the reticent line diagram in Watson and Crick's seminal article[2] (Fig. 2) to its position in today's world of global imagery is extraordinary. It is therefore timely to look at some of the representational issues involved in science communication, and then at a few selected instances of the various guises in which the molecule has replicated itself within varied visual habitats.

A model of communication

Looking back on the laconic article in *Nature* that announced the structure of DNA, which we tend to assume in retrospect provided the definitive solution, it is remarkable how little was actually given away. This is true of the article's sole diagram, drawn by Odile Crick, Francis's wife, which represented the sugar chains as directional ribbons, while the bases were rudimentary rods represented flat on (Fig. 2). Along the vertical axis runs the central pole, depicted as a thick line that is broken where the bases lie in front. This axis is a visually useful point of reference, but its early ubiquity seems to depend on the structural necessities of physical models. The developed model, composed from standard brass components with tailor-made metal bases, provided a more detailed and explicit entity for debate and large-scale publicity, although the famous photographs by Anthony Barrington Brown (Fig. 3), taken for an article in *Time* magazine, were actually staged a few months later.

The model of the double helix — like those of other molecules, such as the model of haemoglobin by Perutz — played an important role in scientific understanding, being both based upon and in turn affecting the acts of scientific conceptualization. Overtaken by more refined models made at King's College London, including the widely illustrated space-filling model with Van der Waals surfaces by Wilkins (Fig. 4), the ramshackle masterpiece of Watson and Crick

passed the way of so many obsolete bits of scientific paraphernalia. When, 23 years after its making, some of the specially cut plates (Fig. 5) resurfaced in Bristol, they were incorporated into a pious reconstruction by Farooq Hussain of King's College. Like an ancient Greek vase reassembled from chards, the semi-original model is now a treasured cultural icon, displayed in the Science Museum in London.

Communicating the complex structure and, in due course, the awesomely intricate behaviour of the modular molecule, has provided an unparalleled challenge for

Figure 3 Anthony Barrington Brown's photograph of Watson and Crick with their model of DNA at the Cavendish Laboratory in Cambridge, 21 May 1953.

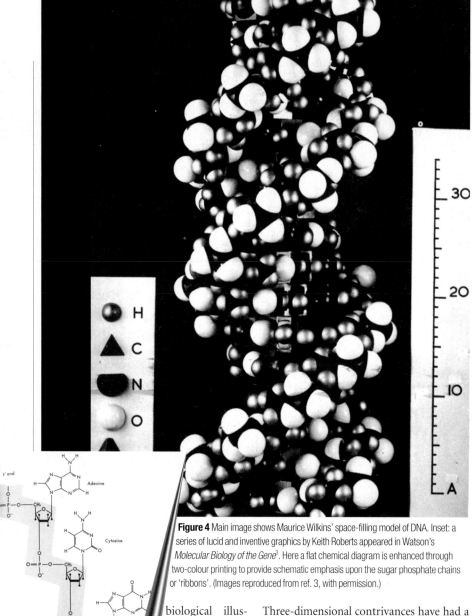

Figure 5 One of the specially cut plates used by Watson and Crick in their model of the structure of DNA.

Figure 6 Cover of *Nature* human genome issue, published on 15 February 2001. The image, by Eric Lander, was created by Runaway Technology Inc. using PhotoMosaic by Robert Silvers from original artwork by Darryl Leja (courtesy of the Whitehead Institute for Biomedical Research). Gregor Mendel, James Watson and Francis Crick are amongst the crowd.

Figure 4 Main image shows Maurice Wilkins' space-filling model of DNA. Inset: a series of lucid and inventive graphics by Keith Roberts appeared in Watson's *Molecular Biology of the Gene*[3]. Here a flat chemical diagram is enhanced through two-colour printing to provide schematic emphasis upon the sugar phosphate chains or 'ribbons'. (Images reproduced from ref. 3, with permission.)

biological illustrators and model-makers. This is vividly shown by Keith Roberts's illustrations in successive editions of Watson's *Molecular Biology of the Gene*, beginning in 1965, which chart the complex interplay between developing science, graphic ingenuity and technologies of reproduction[3] (see Fig. 4, inset).

As the complex functioning of DNA became increasingly elucidated, so methods and conventions of illustration that privileged behaviour over structure played an ever more conspicuous role. As with sets of illustrations in any science, the visual conventions not only reflect what scientists want to show, but also provide an important framework for thinking and visualization in the process of research itself. Subsequently, the resources of computer design, stereoscopy and, in particular, animation have provided a vivid sense of spatial and temporal processes, only partly possible in conventional text and illustration.

Three-dimensional contrivances have had a crucial role from the outset, unsurprisingly given a structure that taxes our powers of spatial visualization. Even in the age of computer graphics, there is still a pedagogic and popular market for kits using a variety of space-filling units.

A number of notable models of DNA and other large molecules have become revered items, typically displayed in protective cases in the foyers of laboratories, where they form part of the visual furniture that speaks of the enterprise of biological science in general and that of the institution in particular. For the sub-species of biologist known as 'molecular', the seductive geometry of DNA helps to underline the fundamental 'hardness' of their science, compared to natural historians and ecologists from whom they have become institutionally distinguished. It is in this spirit, less of didactic instruction than of emblematic signalling, that the double helix has become the icon for the communication of a generalized message. Few have any trouble in recognizing the ghostly twist that emerges from the mosaic of faces on the cover of the *Nature* issue devoted to the human genome on 15 February 2001 (Fig. 6).

Similarly, any hint of the double twist in any logo of a laboratory or biotech company is immediately identifiable.

Aesthetics and meaning

Given the role of aesthetic intuitions in the processes that led to its discovery, and its recognition as 'right', it is understandable that the double helix has itself assumed the guise of a work of art, not least in three-dimensional form. For artists, the attraction of a form that is both beautiful and full of all kinds of scientific and social significance is considerable.

Some grand structures have been commissioned by academic institutions, whereby the artist has basically been given a brief to make a sculpture representing DNA, much like a sculptor might be commissioned to produce an anatomically accurate image of a naked figure. For example, in 1998 Roger Berry produced a huge sculpture hanging down the central well of a multi-story staircase at the University of California at Davis (Fig. 7). Another rendering of the helical structure, *Spirals Time — Time Spirals*, resides on a hillock in the grounds of the Cold Spring Harbor Laboratory (Fig. 8). Designed by the artist, architect and theorist of post-

Figure 7 (left) *Portrait of a DNA Sequence* by Roger Berry (1998) at the Life Sciences Addition building, University of California, Davis. **Figure 8** *Spirals Time — Time Spirals* by Charles Jencks (2000) at Cold Spring Harbor Laboratory.

modernism, Charles Jencks, it stands at the heart of a programme of commissioning and collecting artwork that expresses the vision of Watson — who became director of Cold Spring Harbor Laboratory in 1968 and president in 1994 — of an environment in which the visual stimulation of the surroundings is integral to the conduct of high-level mental activity.

In pursuit of structural aesthetics, the British sculptor, Mark Curtis, proposed a reformed molecular structure for DNA. As an artist concerned with geometrical logic and symmetries, Curtis was worried about the 'ugly' engineering of the Watson–Crick version. Rather than using the sugar phosphate backbones to control the helices, he proposed that stacked base pairs, coupled in an opposite orientation from the accepted bonding, comprises a helix of pentagonal plates around a central void of decagonal cross-section. The geometrical and structural probity of Curtis's models, which eschew a central pole, made it on to a British millennium stamp (Fig. 9), if not into the world of scientific orthodoxy. In a real sense, the molecular biologists' rejection of Curtis's effort to re-design DNA on the basis of *a priori* principles represents an extreme example of the tension within science itself between the polar instincts of the modellers and the empiricists.

Alongside such sculptural exploitations of the inherent beauty of the double helix has run a strand of artistic iconography that has been more overtly concerned with meaning. The tone for the more fantastical exploitations was set by the flamboyant surrealist, Salvador Dali, as ever concerned with the metaphysical potential implicit in scientific imagery. During the late 1950s and 1960s, the DNA molecule features as a symbolic vision, lurking in a surreal hinterland between galactic mystery and spiritual significance (as a kind of Jacob's Ladder). His *Butterfly Landscape, The Great Masturbator in Surrealist Landscape with DNA* (1957–8; Fig. 10) locates a prettified evocation of a space-filling model in one of Dali's typically barren landscapes inhabited by sub-Freudian enigmas, designed to conjure up a dreamworld of obscure sexual fantasy[4]. Subsequent artists, particularly those who have engaged with the social implication of molecular biology and genetic engineering, have located images of DNA in contexts of meaning that are less obscure and more polemic.

This savagely selective glance at DNA art — omitting such contemporary luminaries of genetic art as Suzanne Anker[5] (Fig. 11; www.geneculture.org), David Kremers

Figure 9 Paintings of DNA models on a 'Millennium Collection' stamp, designed by Mark Curtis (1999–2000), from the UK Royal Mail's Scientists' Tale collection.

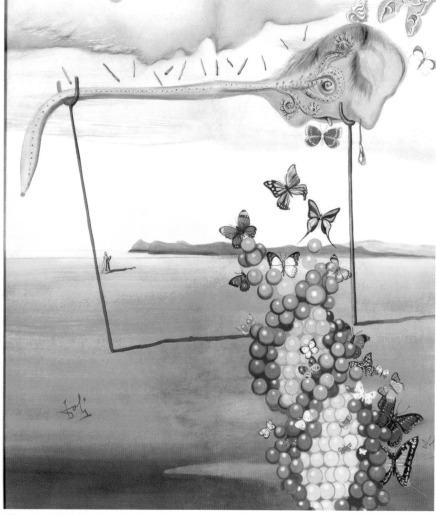

Figure 10 *Butterfly Landscape, The Great Masturbator in Surrealist Landscape with DNA* by Salvador Dali, 1957–8. Private collection.

Figure 11 *Zoosemiotics: Primates, Frog, Gazelle, Fish* (detail) by Suzanne Anker (1993).

(http://davidkremers.caltech.edu/), Ellen Levy (http://www.geneart.org/genome-levy.htm), Sonya Rapoport (http://users.lmi.net/sonyarap/transgenicbagel/) and Gary Schneider (http://www.icp.org/exhibitions/schneider/) — can barely claim to be representative even of the main range of possibilities. In particular, exploitation of the replicating potential of DNA to generate self-organizing images — exemplified by Marc Quinn's genetic portrait of Sir John Sulston from Sulston's own DNA, fragmented and replicated in bacterial colonies on plates of agar jelly[6] — shows that some artists' engagement with DNA is maturing beyond iconographical opportunism.

But as with the *Mona Lisa*, opportunism will always be the name of a prominent public game. Typical of this tendency is the introduction by the perfume company Bijan in 1993 of a fragrance named DNA. Ironically, we learn from the maker's blurb that "DNA is recommended for casual use". Such is the destiny of one of the greatest popular icons. ☐

doi:10.1038/nature01403

1. Coen, E. *The Art of Genes. How Organisms Grow Themselves* (Oxford Univ. Press, Oxford, 1999).
2. Watson, J. D. & Crick, F. H. C. A structure for deoxyribose nucleic acid. *Nature* **171**, 737–738 (1953).
3. Watson, J. D. *Molecular Biology of the Gene* (W. A. Benjamin, Inc., New York, 1965).
4. Descharnes, H. & Neret, G. *Salvador Dali 1904–1989* (Taschen, New York, 1998).
5. Nelkin, D. & Anker, S. The influence of genetics on contemporary art. *Nature Rev. Genet.* **3**, 967–971 (2002).
6. Kemp, M. Reliquary and replication. *A Genomic Portrait: Sir John Sulston* by Marc Quinn. *Nature* **413**, 778 (2001).

Acknowledgements
R. Hodgson provided research assistance. Important advice has been gratefully received from S. de Chadarevian and H. Kamminga of the Department of the History and Philosophy of Science at Cambridge University, E. Levy, E. Coen, and K. Roberts of the John Innes Centre at Norwich.

Original reference: *Nature* **421**, 416–420 (2003).

Box 1
2003 exhibitions celebrating art in the age of the double helix

2003 is to throw up a series of exhibitions, including:

- Representations of the Double Helix, at the Whipple Museum of the History of Science, Department of History and Philosophy of Science, University of Cambridge, UK. The exhibition is curated by Soraya de Chadarevian and Harmke Kamminga, with the assistance of Corrina Bower, and will run from January to December 2003.
- Genetic Expressions: Art After DNA, at the Heckscher Museum of Art, Huntington, New York. Curated by Elizabeth Meryman and Lynn Gamwell, the exhibition will run from 28 June to 7 September 2003.
- From Code to Commodity: Genetics and Visual Art, at the New York Academy of Sciences. The exhibition runs from 13 February to 11 April 2003 and is curated by Dorothy Nelkin and Suzanne Anker. Cold Spring Harbor Laboratory Press are publishing a book by Nelkin and Anker entitled *The Molecular Gaze: Art in the Genetic Age*.
- PhotoGENEsis: Opus 2 — Artists' Response To the Genetic Information Age, at the Santa Barbara Museum of Art, 9 November 2002–9 February 2003.

Portrait of a molecule

Philip Ball

Nature, *The Macmillan Building, 4 Crinan Street, London N1 9XW, UK (e-mail: p.ball@nature.com)*

The double helix is idealized for its aesthetic elegant structure, but the reality of DNA's physical existence is quite different. Most DNA in the cell is compressed into a tangled package that somehow still exposes itself to meticulous gene-regulatory control. Philip Ball holds a mirror up to what we truly know about the mysteries of DNA's life inside a cell.

"Each level of organization represents a threshold where objects, methods and conditions of observation suddenly change… Biology has then to articulate these levels two by two, to cross each threshold and unveil its peculiarities of integration and logic." François Jacob, in *The Logic of Life* (Penguin, London, 1989).

Rather like those of Albert Einstein, DNA's popular images are hardly representative. While it is fashionable in these post-genome days to show it as an endless string of A's, C's, G's and T's, this year's anniversary will surely be replete with two kinds of picture. One shows the famous double helix, delightfully suggesting the twin snakes of Wisdom and Knowledge intertwining around the caduceus, the staff of the medic's god Hermes. The other reveals the X-shaped symbol of inheritance, the chromosome.

But it is rare that DNA looks this good. For only a couple of hours during the early stages of the cell cycle, as the cell prepares to divide, the genome is compacted into its distinctive chromosomal fragments (Fig. 1). The rest of the time you will search the eukaryotic cell in vain for those molecular tetrapods. What you find instead in the cell nucleus is, apparently, a tangled mess.

And don't think that this will, on closer inspection, turn out to be woven from that elegant, pristine double helix. Rather, the threads are chromatin — a filamentary assembly of DNA and proteins — in which only very short stretches of the naked helix are fleetingly revealed. Although chromosomes are often equated with DNA, there is actually about twice as much protein as DNA in chromatin. And about 10 per cent of the mass of a chromosome is made up of RNA chains, newly formed (or in

the act of forming) on the DNA template in the process called transcription.

Zooming in on DNA

If we want to know how DNA really functions, it is not enough to zoom in to the molecular level with its beautifully simple staircase of base pairs. Textbooks tend, understandably, to show replication as the steady progress of the DNA-synthesizing enzyme DNA polymerase along a linear single strand laid out like a railway line (see accompanying article by Alberts, page 117), and RNA polymerase doing likewise in transcription. One has the impression of the genome as a book lying open, waiting to be read.

However, it is not so straightforward. The book is closed up, sealed, and packed away. Moreover, the full story is not merely what is written on the pages; these operations on DNA involve information transmission over

many length scales. Perhaps those who do not routinely have to delve into the intricacies of genome function have acquired such a simplistic picture of it all because, until relatively recently, these length scales were considered largely out of bounds for molecular science. We know about molecules; we know about cells and organelles; but the stuff in between is messy and mysterious.

We speak of molecular biology and cell biology, but no one really talks of mesobiology. Yet that is the level of magnification at which much of the action takes place: the scale of perhaps a few to several hundred nanometres. How DNA is arranged on these scales seems to be central to the processes of replication and transcription that we have come to think of in terms of neat base pairings, yet it is precisely here that our understanding remains the most patchy.

Partly that's because the mesoscale represents, quite literally, a difficult middle ground. It encompasses too many atoms for one to apply straightforward molecular mechanics, with its bond bending and breaking; yet the graininess still matters, the continuum has not yet become a good approximation. As Bustamante and colleagues show elsewhere in this collection (see page 109), looking at DNA on a scale where it flexes and twists like a soft rod reveals how the mechanical and the molecular interact.

Take the problem of supercoiling, for example. The closed loops of bacterial DNA can develop twists like those in a Möbius

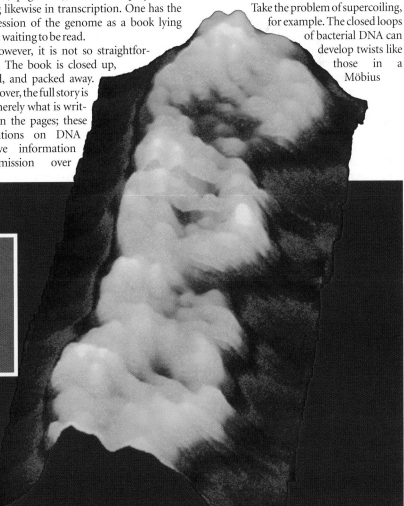

Figure 1 Inset: coloured scanning electron micrograph of a pair of human chromosomes. Main image: scanning tunnelling micrograph showing approximately three turns of a DNA double helix. The image is created by scanning a fine point just above the surface of a DNA molecule and electronically recording the height of the point as it moves across the specimen.

INSET: A. SYRED/SPL. MAIN IMAGE: DRISCOLL, YOUNGQUIST & BALDESCHWIELER, CALTECH/SPL

strip, which either 'overwind' or 'underwind' the helix. Generally there is some degree of underwinding (negative supercoiling) such that there is one negative supercoil for every 200 base pairs (bp) or so. This has an energy cost of around -9 kcal mol^{-1}, which manifests itself in physiological effects. In bacteria, too much supercoiling can inhibit growth, which is why enzymes called topoisomerases exist to release it. On the other hand, negative supercoiling tends to unwind the double helix, which is needed to initiate strand separation for DNA replication.

Although the chromosomal DNA of eukaryotes has free ends, it too is prone to supercoiling, as it seems typically to be attached in large loops to a filamentous structure called the nuclear matrix that coats the inside of the nuclear membrane. The attachment may in fact be necessary for both replication and transcription to take place.

Packaging problem

Stretched into a linear double helix, the three billion or so base pairs of human DNA would measure 1.8 m. This strand, snipped into 46 chromosomes, has to be packed into a nucleus just 6 μm or so across. As a result, the DNA chains are far from the idealized picture of molecules floating in an infinite solvent. They have a density of around 100 mg per ml, comparable to that of a highly viscous polymer gel.

The packing ratio for the chains is therefore enormous. In the smallest human chromosome, a length of DNA 14 mm long is compressed into a chromosome about 2 μm long: a packing ratio of 7,000. The first stage in solving this packaging problem is to wind the DNA around protein disks to form a bead-like nucleosome (see accompanying article by Felsenfeld and Groudine, page 134). Each disk is an octamer of four types of histone protein; a fifth histone, called H1, seals the DNA to the disk at the point where the winding starts and ends. Each nucleosome, 6 nm high by 11 nm in diameter, binds around 200 bp of DNA in two coils, and there is very little 'free' DNA between adjacent nucleosomes: sometimes as little as 8 bp.

The string of nucleosomes forms a fibre about 10 nm thick, which is then packaged into a filament three times as wide. This 30-nm fibre is the basic element of chromatin — yet we still don't know its structure. It is widely held to be composed of nucleosomes arranged in a solenoid, but hard evidence for this is scanty. How many celebrations of the double helix will admit that, 50 years on, we don't really know what DNA at large in the cell looks like?

The 30-nm fibre is further folded and condensed to give a packing ratio of around 1,000 in chromosomes during interphase (the time between cell divisions), and around ten times that in the X-shaped chromosomes of mitosis (cell division). How this happens is even more of a mystery. For mitotic chromosomes it

was thought until only recently that there might be a contiguous protein scaffold holding the whole affair together; but now it seems that the structural integrity must come from chromatin crosslinking[1]. All the histones seem to have higher-order structural functions. Multi-subunit protein complexes in yeast called SWI/SNF and RSC (both of which seem to have human homologues) are chromatin-remodelling machines, which distort histone–DNA contacts or transfer histones between DNA molecules, exposing the DNA to attack by DNA-cleaving enzymes called nucleases. How they work remains hazy[2]. According to one recent study[3], DNA engaged by such complexes 'behaves as if it were free and bound at the same time'. Or in other words, as if 'free' and 'bound' were notions too simplistic to have much meaning here. What is clear is that these chromatin-shaping machines are important in transcription: cells lacking RSC are no longer viable.

There are in fact two types of chromatin in the nucleus of an interphase eukaryotic cell. Euchromatin is the most abundant: it is relatively dispersed, with the tangled-net appearance of a polymer gel. Heterochromatin is much denser (virtually solid-like), comparable to the density of mitotic chromosomes, and is confined to a few small patches. The invitation is to regard euchromatin as 'active' DNA, unpacked enough to let the transcription apparatus get to work on it, while heterochromatin is compressed, like a big data file, until needed. But like just about any other generalization about DNA's structure and behaviour, this one quickly breaks down. Clearly only a fraction of a cell's euchromatin is made up of transcribable DNA in the first place (so why not pack the rest away?); and even chromosomes containing a large amount of heterochromatin can be transcriptionally active. Some researchers think that 'euchromatin' and 'heterochromatin' are just blanket terms for many things we don't understand: further hierarchies of DNA organization yet to be revealed.

Structured chaos

Certainly, there seems to be more to the nucleus than a disorderly mass of DNA. It is a constantly changing structure, but not randomly: there is method in there somewhere. Specific chromosomes occupy discrete nuclear positions during interphase, and these positions can change in a deterministic way in response to changes in the cell's physiological state.

And the euchromatin itself has an internal logic, albeit one only partly decoded. It has been proposed that DNA has sequences called scaffold/matrix-attached regions (S/MARs), recurring typically every 10–100 kbp, that bind to the nuclear matrix to divide up the chromosome into loops[2]. Yet the existence of not only S/MARs, but also the nuclear scaffold itself, has been questioned. There is no sign of the scaffold during mitosis, and the material it is thought to be

composed of may be nothing more than a mess of denatured proteins.

Be that as it may, the organization of the loops seems to be important for compaction of DNA and for the regulation of gene expression, and each loop may act as an independent unit of gene activity. In other words, there is at least one level of superstructural organization in the chromosomes that makes its influence felt at the scale of molecular information transfer. Topoisomerase II is one of several proteins that bind specifically to the putative S/MARs, suggesting that these points are important for controlling supercoiling in the strands.

With all this high-level structure, transcription of DNA is not so much a matter of slotting the parts in place as tugging on the rope. DNA is highly curved around the nucleosomes, the inward-facing groove compressed and the outer one widened. RNA polymerase, at 13 by 14 nm, is about the same size as the nucleosome, yet it binds to a region of DNA around 50 bp long: about a quarter of the entire histone-bound length. So clearly some DNA must leave the surface of the histone core for transcription to proceed. But this core need not be displaced completely. The histone disk actually has a considerable amount of mobility, sometimes described as a corkscrew motion through the DNA coil. The reality is undoubtedly more complex, involving a kind of diffusion of localized defects in the DNA–histone contact.

If all of this destroys the pretty illusion created by the iconic model of Watson and Crick, it surely also opens up a much richer panorama. The fundamental mechanism of information transfer in nucleic acids — complementary base pairing — is so elegant that it risks blinding us to the awesome sophistication of the total process. These molecules do not simply wander up to one another and start talking. They must first be designated for that task, and must then file applications at various higher levels before permission is granted, forming a complex regulatory network (see accompanying article by Hood and Galas, page 130). For those who would like to control these processes, and those who seek to mimic them in artificial systems, the message is that the biological mesoscale, far from being a regime where order and simplicity descend into unpredictable chaos, has its own structures, logic, rules and regulatory mechanisms. This is the next frontier at which we will unfold the continuing story of how DNA works. ☐

doi:10.1038/nature01404

1. Poirier, M. G. & Marko, J. F. Mitotic chromosomes are chromatin networks without a mechanically contiguous protein scaffold. *Proc. Natl Acad. Sci. USA* **99**, 15393–15397 (2002).
2. Van Driel, R. & Otte, A. P. (eds) *Nuclear Organization, Chromatin Structure, and Gene Expression* (Oxford Univ. Press, 1997).
3. Asturias, F. J., Chung, W.-H., Kornberg, R. D. & Lorch, Y. Structural analysis of the RSC chromatin-remodeling complex. *Proc. Natl Acad. Sci. USA* **99**, 13477–13480 (2002).

Original reference: *Nature* **421**, 421–422 (2003).

Ten years of tension: single-molecule DNA mechanics

Carlos Bustamante[*][†], **Zev Bryant**[*] & **Steven B. Smith**[†]

Department of Molecular and Cell Biology, and †Department of Physics and Howard Hughes Medical Institute, University of California, Berkeley, California 94720, USA (e-mail: carlos@alice.berkeley.edu; zev@alice.berkeley.edu; steve@alice.berkeley.edu)

The basic features of DNA were elucidated during the half-century following the discovery of the double helix. But it is only during the past decade that researchers have been able to manipulate single molecules of DNA to make direct measurements of its mechanical properties. These studies have illuminated the nature of interactions between DNA and proteins, the constraints within which the cellular machinery operates, and the forces created by DNA-dependent motors.

The physical properties of the DNA double helix are unlike those of any other natural or synthetic polymer. The molecule's characteristic base stacking and braided architecture lend it unusual stiffness: it takes about 50 times more energy to bend a double-stranded DNA (dsDNA) molecule into a circle than to perform the same operation on single-stranded DNA (ssDNA). Moreover, the phosphates in DNA's backbone make it one of the most highly charged polymers known.

The protein machinery involved in copying, transcribing and packaging DNA has adapted to exploit these unique physical properties (see article by Alberts, pages 117). For example, RNA polymerases (which synthesize RNA from a DNA template) and helicases (which unwind the double helix to provide single-stranded templates for polymerases) have evolved as motors capable of moving along torsionally constrained DNA molecules. DNA-binding proteins can use the polymer's electrostatic potential to cling to DNA while they diffuse along the molecule in search of their target sequences. Topoisomerases break and rejoin the DNA to relieve torsional strain that accumulates ahead of the replication fork.

During the past ten years, direct manipulation of single molecules of DNA has expanded our understanding of the mechanical interactions between DNA and proteins, following a pattern in which basic investigations of DNA elasticity have laid the groundwork for real-time, single-molecule assays of enzyme mechanism.

DNA as a worm-like chain

Although mechanical properties vary according to local sequence and helical structure, the relevant physics of DNA in many biological contexts is usefully described using a coarse-grained treatment such as the worm-like chain (WLC) model[1], which characterizes a polymer using a single parameter, the flexural persistence length (A). The WLC model imagines a polymer as a line that bends smoothly under the influence of random thermal fluctuations. The value of A defines the distance over which the direction of this line persists: correlation between the orientations of two polymer segments falls off exponentially (with decay length A) according to the contour length that separates them. For dsDNA in physiological buffer, $A = \sim 50$ nm.

There is a simple relationship between A and the bending rigidity κ of the polymer represented as an elastic rod[2]: $k_B T A = \kappa$, where k_B is

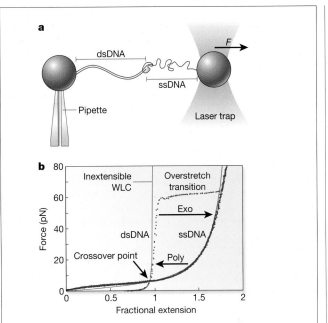

Figure 1 Single-molecule assays of replication[12,13]. **a**, A DNA molecule is stretched between beads held in a micropipette and a force-measuring optical trap[12]. The measured extension is the sum of contributions from the single-stranded DNA (ssDNA) and double-stranded DNA (dsDNA) segments. **b**, Force versus extension for dsDNA and ssDNA molecules, obtained in the instrument in panel **a**. Arrows show changes in extension observed at constant tension during polymerization (Poly) or force-induced exonuclease activity (Exo).

Boltzmann's constant and T is the temperature. According to this relation, the energy required to bend a segment of DNA of length L through an angle θ and a radius of curvature R/L is:

$$E(\theta) = \frac{k_B T A L}{2R^2} = \frac{k_B T A}{2L}\theta^2$$

This model, therefore, predicts that it is energetically more favourable to bend the molecule smoothly, spreading the strain over large distances, than to bend it sharply at discrete locations. This mechanical property is central to interactions with regulatory proteins that bend DNA severely upon binding. The biological relevance of these bends is demonstrated by the enhancement of DNA recombination and gene transcription observed when specific protein-binding sites for activators are replaced by intrinsically bent DNA sequences[3] or by binding sequences for unrelated DNA-bending proteins in the presence of these proteins[4].

To bend DNA, proteins must convert part of their binding energy into mechanical work, as illustrated by an experiment in which a binding sequence was pre-bent towards the major groove by placing it in a DNA minicircle. The affinity of a transcription factor (TBP) for this binding site was found to be 300-fold higher (equivalent to a free-energy change of 3.4 kcal mol⁻¹) when the sequence was pre-bent in the same direction as TBP-induced bending, relative to pre-bending in the opposite direction[5]. This increase can largely be accounted for by the difference in bending energy between the two initial DNA conformations, which by the equation above is predicted to be 3.2 kcal mol⁻¹.

The high linear charge density of the double helix provides one mechanism for converting binding energy into work. DNA's structure is pre-stressed by electrostatic self-repulsion, as a result of the negatively charged phosphate backbone of the double helix. Therefore, asymmetric neutralization of the DNA helix (for example, by a DNA-binding protein that presents a positively charged face) can lead to compression and bending of DNA towards the neutralized face. This

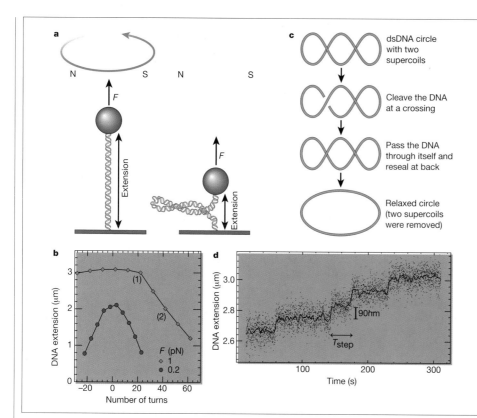

Figure 2 The elastic behaviour of supercoiled DNA molecules[49] forms the basis of single-molecule topoisomerase assays[15–17]. **a**, Molecules are stretched and twisted in magnetic tweezers. Under sufficient torsional strain, a twisted DNA molecule buckles to form plectonemes, shortening the measured extension. **b**, Extension as a function of turns introduced into the molecule remains nearly constant until the buckling transition is reached (1), after which the molecule contracts linearly (2). Underwinding at 1 pN does not cause reduced extension because the strained molecule forms alternative underwound structures (for example, melted DNA) in preference to buckling. **c**, The accepted model of type II topoisomerase action: the enzyme binds to supercoiled DNA, cleaves both strands, and passes the double helix through itself, leading to the removal of two turns. (Redrawn from ref. 50, with permission.) **d**, When topoisomerase II is added to a plectonemed molecule in the magnetic tweezers, 90-nm steps in extension are seen, corresponding to the removal of two turns[15], as predicted by the model.

effect has been elegantly demonstrated by incorporating neutral phosphate analogues or tethered cations onto one face of a DNA molecule[6].

DNA elasticity

The bending elasticity of DNA has consequences beyond short-range interactions with proteins: the WLC model explicitly connects local bending mechanics with the statistics of global conformations. Thus, a polymer with smaller bending rigidity tends to adopt a more compact random-coil structure.

This preference is reflected in the phenomenon of entropic elasticity, which is responsible for the elastic properties of common polymeric materials such as rubber[7]. A flexible polymer coils randomly in solution, resulting in an average end-to-end distance much shorter than its contour length. Pulling the molecule into a more extended chain is entropically unfavourable, as there are fewer possible conformations at longer extensions, with only a single possible conformation (a perfectly straight line) for maximum extension. The resulting entropic force increases as a random coil is pulled from the ends; tensions on the order of $k_B T/A$ (~0.1 pN for dsDNA or 5 pN for ssDNA) are required to extend the molecule significantly.

Direct measurements of force and extension on single molecules of DNA provide the most rigorous test to date of theories of entropic elasticity. When magnets and fluid flow[8] and later optical traps[9,10] were used to stretch DNA molecules attached to micron-scale beads (Fig. 1), the entropic force–extension behaviour of dsDNA was found to agree closely with the WLC model[11]. Tensions of ~6 pN, within the range of forces exerted by characterized molecular motors, stretch dsDNA to ~95% of its contour length.

The intrinsic flexibility of ssDNA causes it to maintain very compact conformations, so that its extension per base pair is shorter than that of dsDNA for forces smaller than ~6 pN. At higher forces, however, the situation is reversed. A single strand is not constrained to follow a helical path, so it becomes nearly twice as long as dsDNA as it is pulled straight (Fig. 1).

From elasticity to enzymology

A quantitative appreciation of the different elastic properties of ssDNA and dsDNA has allowed researchers to observe replication of single DNA molecules[12,13]. In these studies, a molecule of ssDNA was stretched between two surfaces, and a DNA polymerase was allowed to replicate the stretched molecule at a given constant tension. As ssDNA was converted into dsDNA by the polymerase, replication could be followed in real time by monitoring the extension (below 6 pN) or contraction (above 6 pN) of the molecule (Fig. 1).

These studies showed that the rate-limiting step of DNA replication, which involves closing a structural 'fingers' domain of the enzyme, is sensitive to DNA tension and is capable of generating forces as high as 35 pN. Small forces can accelerate the enzyme's activity, probably by helping it to stretch the proximal collapsed template strand into the correct geometry for polymerization. A surprising result was the induction of a strong exonuclease activity (removal of nucleotides) at tensions above 40 pN (ref. 12). This effect provides a novel assay to investigate the proofreading mechanism of DNA polymerases.

Studies of the force–extension behaviour of single supercoiled DNA molecules further illustrate the progression from elasticity measurements to enzymology (Fig. 2). DNA tethers were stretched between a surface and a magnetic bead that could be rotated using magnets[14]. Because the molecule was attached at each end through multiple linkages on both strands, rotation of the bead led to the build up of torsional strain in the molecule. Under tension, such a molecule behaves roughly like a twisted rubber tube: as turns are added, the extension remains nearly constant until a critical amount of torque accumulates and the tube buckles, trading twist for writhe to form plectonemes (units of supercoiled DNA that project out of the molecular axis), thus reducing its apparent extension with each subsequent turn. As tension is increased, so does the energetic penalty for buckling; therefore, more turns must be added to reach the buckling transition.

The activity of topoisomerase II, an enzyme that relaxes supercoils in eukaryotic cells, has been analysed on single plectonemed DNA tethers[15]. Under conditions of limiting ATP, discrete steps in extension were observed that were attributable to single enzymatic turnovers. The size of these steps corresponded to the removal of two turns, confirming the

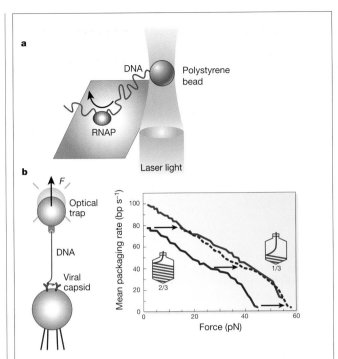

Figure 3 Force generation by DNA-dependent motors. **a**, Transcription. A surface-bound RNA polymerase (RNAP) transcribes against a force exerted by an optical trap[24]. Its velocity remains unchanged against forces up to ~20 pN (implying that translocation is not rate-limiting), and then falls steeply with increasing force. **b**, Bacteriophage DNA packaging. As DNA is pulled in by the φ29 portal motor to fill the viral capsid, the extension of the external DNA becomes shorter. The force–velocity relation for packaging when the capsid is two-thirds full (solid blue) must be shifted by ~15 pN (dashed blue) to match the curve for one-third-full capsids (red), implying the presence of an additional internal force that builds up during DNA packaging, due to highly compressed DNA[32].

accepted model of topoisomerase II action in which the double helix is passed through itself, changing the linking number by two (Fig. 2). Later experiments applied the same methodology to bacterial topoisomerases I and IV (refs 16 and 17, respectively).

Single-molecule assays of topoisomerase IV (ref. 17) helped reveal that the enzyme has a chiral substrate specificity: it relaxes overwound DNA substantially more efficiently than underwound DNA. This allows the enzyme to relax the positive supercoils formed during replication while avoiding counterproductive relaxation of the negative supercoils present in non-replicating DNA.

These studies also showed that topoisomerase IV relaxes DNA (of either handedness) an order of magnitude faster than had been estimated from bulk studies, helping to resolve a dilemma in which the enzyme's apparent low turnover rate *in vitro* had seemed to be at odds with its demonstrated ability to counteract rapid supercoil formation at the replication fork *in vivo*. This result reflects a general caveat for bulk enzyme kinetics: the presence of inactive enzymes in solution can lead to gross underestimation of the turnover rate per enzyme. Single-molecule assays, including those based on DNA manipulation, can sidestep this issue by selecting only the active fraction for analysis.

DNA unzipped
DNA helicases must generate force to unzip the parental strands during replication (see article by Alberts, page 117). Mechanical unzipping forces for dsDNA were first measured by attaching one strand to a surface (through a dsDNA linker) and pulling on the other strand using a glass needle[18] or optical tweezers[19] as a force transducer. DNA from a bacterial virus, called lambda phage, was unzipped (and re-zipped) at forces between 10 and 15 pN, depend-

ing on the local sequence. The pattern of force fluctuations during unzipping of a particular sequence was remarkably reproducible, and could be rationalized from a simple model incorporating the known difference in stability between adenine–thymine and guanine–cytosine base pairs.

Future studies might use this experimental geometry to investigate helicase function directly. Such an experiment could provide insight into the rates, processivity, force generation and sequence dependence of helicases, complementing the results of previous single-molecule helicase assays which observed translocation without measuring or applying forces[20,21].

In a new application of mechanical unzipping[22], the extra force needed to separate the DNA strands past DNA-binding proteins has been used to map the positions of target sequences, and (by noting the fraction of occupied sites as a function of protein concentration) to measure dissociation constants.

Forces in DNA transcription and packaging
The ability to apply forces on DNA has altered the way we think about DNA-dependent enzymes, by revealing these enzymes to be powerful motors (Fig. 3). Optical tweezers have been used to follow transcription by *Escherichia coli* RNA polymerase against external loads[23–25]. This enzyme can generate forces exceeding those of cytoskeletal motors that drive transport processes within the cell[26]. Its velocity remains unchanged against forces of up to ~20 pN (refs 24,25), showing that the translocation step (which must by definition be force sensitive) is not rate limiting.

An external load can, however, affect the tendency of an enzyme[25] to pause or arrest during transcription. The application of force in an 'aiding' direction[27] reduces pausing and arrest probabilities, presumably by preventing the polymerase from sliding backwards along the template during entry into these inactive states[28]. The same 'backsliding' phenomenon may be responsible for the steep drop in transcriptional velocity seen as opposing force is increased above ~20 pN (ref. 24).

In eukaryotes, forces generated by RNA polymerase or by chromatin-remodelling enzymes might help to displace nucleosomes that would otherwise impede transcription. In support of this idea, several groups have pulled on single chromatin fibres and found that nucleosomes can be removed from DNA by applying a tension of ~20 pN (refs 29–31). At lower forces (~6 pN), reversible modifications of the chromatin structure (such as partial DNA release[31] or disruption of internucleosomal interactions[29]) are observed. These tension-inducible structural rearrangements might be exploited by RNA polymerase or other cellular factors to modulate access of the transcriptional apparatus to chromosomal DNA.

The machine that packs DNA into the viral capsid of the bacteriophage φ29 (a virus that infects bacteria) generates higher forces than have been seen so far for any other translocating (displacing) molecular motor[32] (Fig. 3). A comparison of the force–velocity relation of the motor when the capsid is mostly empty with that when it is nearly full of DNA revealed the presence of a large internal force (up to ~50 pN) pushing back on the motor. This pressure, which must be overcome in order to package the viral genome, presumably arises from the combined effects of configurational entropy loss, elastic bending energy, electrostatic self-repulsion, and changes in hydration of the DNA upon packaging. The potential energy thus stored by the motor in the form of a pre-compressed 'spring' should provide some of the driving force for DNA ejection into the bacterial cytoplasm when the virus infects.

Extreme forms of DNA
The helical structure of DNA is highly adaptable and can assume various forms[33]. Although the helix of dsDNA is typically right-handed and extended in aqueous solution (B form), it can become shorter and wider (A form) in dehydrating solution. Molecules with specific base sequences (alternating purines and pyrimidines) easily assume the left-handed Z form, which is longer than B form and has reverse twist. Recently, single-molecule manipulation experiments

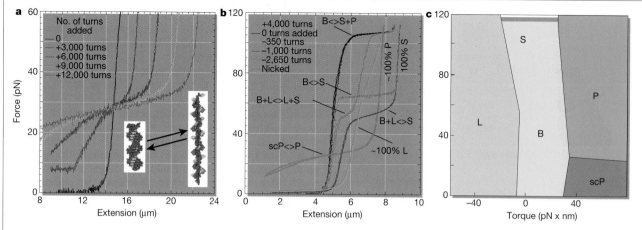

Figure 4 Mechanically induced structural transitions in twist-constrained DNA. **a**, Force–extension curves for a 44.4-kilobase DNA molecule overwound by successively larger numbers of turns (Z.B., M. D. Stone, S.B.S., N. R. Cozzarelli and C.B., unpublished data). As was seen in ref. 39, the high-force curves can by interpreted as the sum of contributions from a fraction of the molecule that remains in B form and a progressively larger fraction that is converted to 'P form', whose structure (proposed from molecular mechanics modelling[39]) is shown in the inset (courtesy of R. Lavery). The curves cross at an 'isosbestic point' marking the force at which B-DNA and P-DNA have equal extensions. **b**, Multiple plateaux occur in force–extension curves for torsionally constrained DNA[35]. (Data shown are for a 14.8-kilobase molecule twisted and stretched in 100 mM Na⁺; Z.B., M. D. Stone, S.B.S., N. R. Cozzarelli and C.B., unpublished data.) **c**, These multiple plateaux can be explained by a 'phase diagram' for DNA under torque and tension (adapted from ref. 41). Coloured regions represent conditions under which pure phases occur; lines indicate conditions for phase coexistence within a molecule. S, overstretched; P, Pauling, sc, supercoiled (shortened by forming plectonemes). L is used here in place of 'Z'[35,41] to denote a phase with an average left-handed twist. Other studies have concluded that this form contains exposed bases, consistent with melted DNA[49]; a mixture of non-canonical forms may in fact be present. A nicked DNA molecule (red curve in **b**) remains at zero torque and therefore crosses the B–S coexistence line at 65 pN.

have revealed the existence of additional helical forms of DNA stabilized by external forces and torques (Fig. 4).

When tension in a nicked DNA molecule is increased to 65 pN, it displays a reversible, cooperative transition to an extended form that is ~70 per cent longer than normal B-DNA[9,34] and with substantially reduced twist[35] (Figs 1b, 4b). But, what is the form of this overstretched dsDNA? Do the strands associate in some specific base-paired structure, dubbed 'S form'[9,34], or does overstretched DNA simply comprise two independent strands of ssDNA[36]? Evidence exists for both models, so the question remains open; a further challenge in single-molecule mechanics is the development of methods to probe the high-resolution structure of manipulated molecules[37].

Twisting of stretched DNA can lead to other structural transitions[35,38,39]. For example, after a critical amount of overwinding has been introduced into a molecule (Fig. 4a), it gets progressively longer with additional twisting, implying cooperative conversion to an overextended form with greatly increased helicity (~2.6 base pairs per turn, compared with 10.5 base pairs per turn for B-DNA). The evidence[39] suggests an inside-out double helix reminiscent of the structure proposed by Linus Pauling in 1953 (ref. 40) and therefore dubbed Pauling DNA (P-DNA).

Complex force–extension curves with multiple force plateaux are seen when single DNA molecules are twisted in either direction and pulled to high forces (ref. 35 and Fig. 4b). A simple model to account for these features assumes that DNA has five interconvertible structural forms[41]. This model predicts a force–torque 'phase diagram'(Fig. 4c), thus framing mechanically induced structural transitions in terms of coexistence lines, critical stresses, and triple-points. Such a model might be tested by direct measurements of torque on stretched and twisted DNA — so far, this quantity has been inferred only indirectly from force–extension experiments. It remains to be determined whether molecular motors can generate sufficient concomitant torque and tension to generate 'extreme forms' of DNA in a biological context.

From mechanics to nanotechnology

Single-molecule manipulation of DNA has illuminated the mechanical basis of interactions between DNA and the molecular machinery involved in transcription, replication and recombination. Over the next decade, these studies are likely to expand to include detailed analyses of the mechanical interactions of many factors involved in these fundamental cellular processes. Because of the potentially large class of motors that track the DNA helix (as demonstrated elegantly for RNA polymerase[42]), necessary technical improvements will include direct measurement of torque in experiments that decouple twisting from bending.

Outside of traditional DNA biology, the ease of synthesis and well-characterized elasticity of DNA make it an ideal material for stiff molecular 'handles' to manipulate other molecules. So far, such handles have been used to mechanically unfold molecules of RNA[43], but covalent attachment of DNA segments to protein molecules has also been demonstrated[44], opening the door for the next generation of forced protein (un)folding studies[45] and perhaps mechanical assays of domain motion in enzymes.

Engineers have recently exploited the properties of DNA to construct self-assembled nanomachines, such as artificial DNA-based devices driven by strand displacement[46,47] or chemically induced structural rearrangements (ref. 48; and see article by Seeman, page 113). DNA micromanipulation techniques will help assess the utility of this new class of molecular machines for which force and torque generation have yet to be measured. The past decade has provided a new perspective of the mechanical nature of the double helix. The next decade promises deeper insight into its interactions with the cellular machinery and its potential for constructing sophisticated nanomachines.

doi:10.1038/nature01405

1. Kratky, O. & Porod, G. Röntgenuntersushung gelöster Fagenmoleküle. *Rec. Trav. Chim. Pays-Bas* **68**, 1106–1123 (1949).
2. Schellman, J. A. Flexibility of DNA. *Biopolymers* **13**, 217–226 (1974).
3. Goodman, S. D. & Nash, H. D. Functional replacement of a protein-induced bend in a DNA recombination site. *Nature* **341**, 251–254 (1989).
4. Perez-Martin, J. & Espinosa, M. Protein-induced bending as a transcriptional switch. *Science* **260**, 805–807 (1993).
5. Parvin, J. D., McCormick, R. J., Sharp, P. A. & Fisher, D. E. Pre-bending of a promoter sequence enhances affinity for the TATA-binding factor. *Nature* **373**, 724–727 (1995).
6. Strauss, J. K. & Maher, L. J. III DNA bending by asymmetric phosphate neutralization. *Science* **266**, 1829–1834 (1994).
7. Beuche, F. *Physical Properties of Polymers* (Interscience, New York, 1962).

8. Smith, S. B., Finzi, L. & Bustamante, C. Direct mechanical measurements of the elasticity of single DNA molecules by using magnetic beads. *Science* **258**, 1122–1126 (1992).

9. Smith, S. B., Cui, Y. & Bustamante, C. Overstretching B-DNA: the elastic response of individual double-stranded and single-stranded DNA molecules. *Science* **271**, 795–799 (1996).

10. Wang, M. D., Yin, H., Landick, R., Gelles, J. & Block, S. M. Stretching DNA with optical tweezers. *Biophys. J.* **72**, 1335–1346 (1997).

11. Bustamante, C., Marko, J. F., Siggia, E. D. & Smith, S. Entropic elasticity of lambda-phage DNA. *Science* **265**, 1599–1600 (1994).

12. Wuite, G. J., Smith, S. B., Young, M., Keller, D. & Bustamante, C. Single-molecule studies of the effect of template tension on T7 DNA polymerase activity. *Nature* **404**, 103–106 (2000).

13. Maier, B., Bensimon, D. & Croquette, V. Replication by a single DNA polymerase of a stretched single-stranded DNA. *Proc. Natl Acad. Sci. USA* **97**, 12002–12007 (2000).

14. Strick, T. R., Allemand, J. F., Bensimon, D., Bensimon, A. & Croquette, V. The elasticity of a single supercoiled DNA molecule. *Science* **271**, 1835–1837 (1996).

15. Strick, T. R., Croquette, V. & Bensimon, D. Single-molecule analysis of DNA uncoiling by a type II topoisomerase. *Nature* **404**, 901–904 (2000).

16. Dekker, N. H. *et al.* The mechanism of type IA topoisomerases. *Proc. Natl Acad. Sci. USA* **99**, 12126–12131 (2002).

17. Crisona, N. J., Strick, T. R., Bensimon, D., Croquette, V. & Cozzarelli, N. R. Preferential relaxation of positively supercoiled DNA by *E. coli* topoisomerase IV in single-molecule and ensemble measurements. *Genes Dev.* **14**, 2881–2892 (2000).

18. Essevaz-Roulet, B., Bockelmann, U. & Heslot, F. Mechanical separation of the complementary strands of DNA. *Proc. Natl Acad. Sci. USA* **94**, 11935–11940 (1997).

19. Bockelmann, U., Thomen, P., Essevaz-Roulet, B., Viasnoff, V. & Heslot, F. Unzipping DNA with optical tweezers: high sequence sensitivity and force flips. *Biophys. J.* **82**, 1537–1553 (2002).

20. Bianco, P. R. *et al.* Processive translocation and DNA unwinding by individual RecBCD enzyme molecules. *Nature* **409**, 374–378 (2001).

21. Dohoney, K. M. & Gelles, J. χ-Sequence recognition and DNA translocation by single RecBCD helicase/nuclease molecules. *Nature* **409**, 370–374 (2001).

22. Koch, S. J., Shundrovsky, A., Jantzen, B. C. & Wang, M. D. Probing protein-DNA interactions by unzipping a single DNA double helix. *Biophys. J.* **83**, 1098–1105 (2002).

23. Yin, H. *et al.* Transcription against an applied force. *Science* **270**, 1653–1657 (1995).

24. Wang, M. D. *et al.* Force and velocity measured for single molecules of RNA polymerase. *Science* **282**, 902–907 (1998).

25. Davenport, R. J., Wuite, G. J., Landick, R. & Bustamante, C. Single-molecule study of transcriptional pausing and arrest by *E. coli* RNA polymerase. *Science* **287**, 2497–2500 (2000).

26. Vale, R. D. & Milligan, R. A. The way things move: looking under the hood of molecular motor proteins. *Science* **288**, 88–95 (2000) .

27. Forde, N. R., Izhaky, D., Woodcock, G. R., Wuite, G. J. & Bustamante, C. Using mechanical force to probe the mechanism of pausing and arrest during continuous elongation by *Escherichia coli* RNA polymerase. *Proc. Natl Acad. Sci. USA* **99**, 11682–11687 (2002).

28. Landick, R. RNA polymerase slides home: pause and termination site recognition. *Cell* **88**, 741–744 (1997).

29. Cui, Y. & Bustamante, C. Pulling a single chromatin fiber reveals the forces that maintain its higher-order structure. *Proc. Natl Acad. Sci. USA* **97**, 127–132 (2000).

30. Bennink, M. L. *et al.* Unfolding individual nucleosomes by stretching single chromatin fibers with optical tweezers. *Nature Struct. Biol.* **8**, 606–610 (2001).

31. Brower-Toland, B. D. *et al.* Mechanical disruption of individual nucleosomes reveals a reversible multistage release of DNA. *Proc. Natl Acad. Sci. USA* **99**, 1960–1965 (2002).

32. Smith, D. E. *et al.* The bacteriophage φ29 portal motor can package DNA against a large internal force. *Nature* **413**, 748–752 (2001).

33. Calladine, C. R. & Drew, H. *Understanding DNA* (Academic, London, 1997).

34. Cluzel, P. *et al.* DNA: an extensible molecule. *Science* **271**, 792–794 (1996).

35. Leger, J. F. *et al.* Structural transitions of a twisted and stretched DNA molecule. *Phys. Rev. Lett.* **83**, 1066–1069 (1999).

36. Williams, M. C., Rouzina, I. & Bloomfield, V. A. Thermodynamics of DNA interactions from single molecule stretching experiments. *Acc. Chem. Res.* **35**, 159–166 (2002).

37. Wilkins, M. H. F., Gosling, R. G. & Seeds, W. E. Nucleic acid: an extensible molecule? *Nature* **167**, 759–760 (1951).

38. Strick, T. R., Allemand, J. F., Bensimon, D. & Croquette, V. Behavior of supercoiled DNA. *Biophys. J.* **74**, 2016–2028 (1998).

39. Allemand, J. F., Bensimon, D., Lavery, R. & Croquette, V. Stretched and overwound DNA forms a Pauling-like structure with exposed bases. *Proc. Natl Acad. Sci. USA* **95**, 14152–14157 (1998).

40. Pauling, L. & Corey, R. B. A proposed structure for the nucleic acids. *Proc. Natl Acad. Sci. USA* **39**, 84–97 (1953).

41. Sarkar, A., Leger, J. F., Chatenay, D. & Marko, J. F. Structural transitions in DNA driven by external force and torque. *Phys. Rev. E* **63**, 051903-1–051903-10 (2001).

42. Harada, Y. *et al.* Direct observation of DNA rotation during transcription by *Escherichia coli* RNA polymerase. *Nature* **409**, 113–115 (2001).

43. Liphardt, J., Onoa, B., Smith, S. B., Tinoco, I. J. & Bustamante, C. Reversible unfolding of single RNA molecules by mechanical force. *Science* **292**, 733–737 (2001).

44. Hegner, M. DNA Handles for single molecule experiments. *Single Mol.* **1**, 139–144 (2000).

45. Carrion-Vazquez, M. *et al.* Mechanical design of proteins studied by single-molecule force spectroscopy and protein engineering. *Prog. Biophys. Mol. Biol.* **74**, 63–91 (2000).

46. Yurke, B., Turberfield, A. J., Mills, A. P., Simmel, F. C. & Neumann, J. L. A DNA-fuelled molecular machine made of DNA. *Nature* **406**, 605–608 (2000).

47. Yan, H., Zhang, X., Shen, Z. & Seeman, N. C. A robust DNA mechanical device controlled by hybridization topology. *Nature* **415**, 62–65 (2002).

48. Mao, C., Sun, W., Shen, Z. & Seeman, N. C. A nanomechanical device based on the B–Z transition of DNA. *Nature* **397**, 144-146 (1999).

49. Strick, T., Allemand, J., Croquette, V. & Bensimon, D. Twisting and stretching single DNA molecules. *Prog. Biophys. Mol. Biol.* **74**, 115–140 (2000).

50. Alberts, B. *et al. Molecular Biology of the Cell* (Garland, New York, 2002).

Original reference: *Nature* **421**, 423–427 (2003).

DNA in a material world

Nadrian C. Seeman

Department of Chemistry, New York University, New York 10003, USA
(e-mail: ned.seeman@nyu.edu)

The specific bonding of DNA base pairs provides the chemical foundation for genetics. This powerful molecular recognition system can be used in nanotechnology to direct the assembly of highly structured materials with specific nanoscale features, as well as in DNA computation to process complex information. The exploitation of DNA for material purposes presents a new chapter in the history of the molecule.

"The nucleic-acid 'system' that operates in terrestrial life is optimized (through evolution) chemistry incarnate. Why not use it ... to allow human beings to sculpt something new, perhaps beautiful, perhaps useful, certainly unnatural." Roald Hoffmann, writing in *American Scientist*, 1994 (ref. 1).

The DNA molecule has appealing features for use in nanotechnology: its minuscule size, with a diameter of about 2 nanometres, its short structural repeat (helical pitch) of about 3.4–3.6 nm, and its 'stiffness', with a persistence length (a measure of stiffness) of around 50 nm. There are two basic types of nanotechnological construction: 'top-down' systems are where microscopic manipulations of small numbers of atoms or molecules fashion elegant patterns (for example, see ref. 2), while in 'bottom-up' constructions, many molecules self-assemble in parallel steps, as a function of their molecular recognition properties. As a chemically based assembly system, DNA will be a key player in bottom-up nanotechnology.

The origins of this approach date to the early 1970s, when *in vitro* genetic manipulation was first performed by tacking together molecules with 'sticky ends'. A sticky end is a short single-stranded overhang protruding from the end of a double-stranded helical DNA molecule. Like flaps of Velcro, two molecules with complementary sticky ends — that is, their sticky ends have complementary arrangements of the nucleotide bases adenine, cytosine, guanine and thymine — will cohere to form a molecular complex.

Sticky-ended cohesion is arguably the best example of programmable molecular recognition: there is significant diversity to possible sticky ends (4^N for N-base sticky ends), and the product formed at the site of this cohesion is the classic DNA double helix. Likewise, the convenience of solid support-based DNA synthesis[3] makes it is easy to program diverse sequences of sticky ends. Thus, sticky ends offer both predictable control of intermolecular associations and predictable geometry at the point of cohesion. Perhaps one could get similar affinity properties from antibodies and antigens, but, in contrast to DNA sticky ends, the relative three-dimensional orientation of the antibody and the antigen would need to be determined for every new pair. The nucleic acids seem to be unique in this regard, providing a tractable, diverse and programmable system with remarkable control over intermolecular interactions, coupled with known structures for their complexes.

Branched DNA

There is, however, a catch; the axes of DNA double helices are unbranched lines. Joining DNA molecules by sticky ends can yield longer lines, perhaps with specific components in a particular linear

Figure 1 Assembly of branched DNA molecules.
a, Self-assembly of branched DNA molecules into a two-dimensional crystal. A DNA branched junction forms from four DNA strands; those strands coloured green and blue have complementary sticky-end overhangs labelled H and H′, respectively, whereas those coloured pink and red have complementary overhangs V and V′, respectively. A number of DNA branched junctions cohere based on the orientation of their complementary sticky ends, forming a square-like unit with unpaired sticky ends on the outside, so more units could be added to produce a two-dimensional crystal. **b**, Ligated DNA molecules form interconnected rings to create a cube-like structure. The structure consists of six cyclic interlocked single strands, each linked twice to its four neighbours, because each edge contains two turns of the DNA double helix. For example, the front red strand is linked to the green strand on the right, the light blue strand on the top, the magenta strand on the left, and the dark blue strand on the bottom. It is linked only indirectly to the yellow strand at the rear.

or cyclic order in one dimension. Indeed, the chromosomes packed inside cells exist as just such one-dimensional arrays. But to produce interesting materials from DNA, synthesis is required in multiple dimensions and, for this purpose, branched DNA is required.

Branched DNA occurs naturally in living systems, as ephemeral intermediates formed when chromosomes exchange information during meiosis, the type of cell division that generates the sex cells (eggs and sperm). Prior to cell division, homologous chromosomes pair, and the aligned strands of DNA break and literally cross over one another, forming structures called Holliday junctions. This exchange of adjacent sequences by homologous chromosomes — a process called recombination — during the formation of sex cells passes genetic diversity onto the next generation.

The Holliday junction contains four DNA strands (each member of a pair of aligned homologous chromosomes is composed of two DNA strands) bound together to form four double-helical arms flanking a branch point (Fig. 1a). The branch point can relocate throughout the molecule, by virtue of the homologous sequences. In contrast, synthetic DNA complexes can be designed to have fixed branch points containing between three and at least eight arms[4,5]. Thus, the prescription for using DNA as the basis for complex materials with nanoscale features is simple: take synthetic branched DNA molecules with programmed sticky ends, and get them to self-assemble into the desired structure, which may be a closed object or a crystalline array (Fig. 1a).

Other modes of nucleic acid interaction aside from sticky ends are available. For example, Tecto-RNA molecules[6], held together by loop–loop interactions, or paranemic crossover (PX) DNA, where cohesion derives from pairing of alternate half turns in inter-wrapped double helices[7]. These new binding modes represent programmable cohesive interactions between cyclic single-stranded molecules that do not require cleavage to expose bases to pair molecules together. Nevertheless, cohesion using sticky ends remains the most prominent intermolecular interaction in structural DNA nanotechnology.

DNA constructions

It is over a decade since the construction of the first artificial DNA structure, a stick-cube, whose edges are double helices[8] (Fig. 1b). More complex polyhedra and topological constructs[9], such as knots and Borromean rings (consisting of three intricately interlinked circles), followed. But the apparent floppiness of individual branched junctions led to a hiatus before the next logical step: self-assembly into two-dimensional arrays.

This step required a stiffer motif, as it was difficult to build a periodic well-structured array with marshmallow-like components,

Figure 2 Two-dimensional DNA arrays. **a**, Schematic drawings of DNA double crossover (DX) units. In the meiotic DX recombination intermediate, labelled MDX, a pair of homologous chromosomes, each consisting of two DNA strands, align and cross over in order to swap equivalent portions of genetic information; 'HJ' indicates the Holliday junctions. The structure of an analogue unit (ADX), used as a tiling unit in the construction of DNA two-dimensional arrays, comprises two red strands, two blue crossover strands and a central green crossover strand. **b**, The strand structure and base pairing of the analogue ADX molecule, labelled A, and a variant, labelled B*. B* contains an extra DNA domain extending from the central green strand that, in practice, protrudes roughly perpendicular to the plane of the rest of the DX molecule. **c**, Schematic representations of A and B* where the perpendicular domain of B* is represented as a blue circle. The complementary ends of the ADX molecules are represented as geometrical shapes to illustrate how they fit together when they self-assemble. The dimensions of the resulting tiles are about 4 × 16 nm and are joined together so that the B* protrusions lie about 32 nm apart. **d**, The B* protrusions are visible as 'stripes' in tiled DNA arrays under an atomic force microscope.

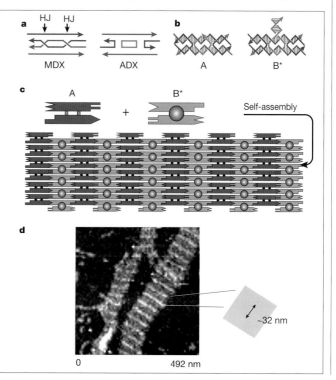

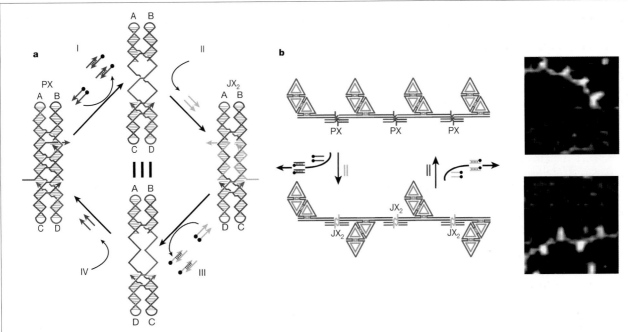

Figure 3 A rotary DNA nanomachine. **a**, The device works by producing two different conformations, depending on which of two pairs of strands (called 'set' strands) binds to the device framework. The device framework consists of two DNA strands (red and blue) whose top and bottom double helices are each connected by single strands. Thus, they form two rigid arms with a flexible hinge in between and the loose ends of the two strands dangling freely. The two states of the device, PX (left) and JX$_2$ (right), differ by a half turn in the relative orientations of their bottom helices (C and D on the left, D and C on the right). The difference between the two states is analogous to two adjacent fingers extended, parallel to each other (right), or crossed (left). The states are set by the presence of green or yellow set strands, which bind to the frame in different ways to produce different conformations. The set strands have extensions that enable their removal when complementary strands are added (steps I and III). When one type of set strand is removed, the device is free to bind the other set strands and switch to a different state (steps II and IV). **b**, The PX–JX$_2$ device can be used to connect 20-nm DNA trapezoid constructs. In the PX state, they are in a parallel conformation, but in the JX$_2$ state, they are in a zig-zag conformation, which can be visualized on the right by atomic force microscopy.

even with a well-defined blueprint (sticky-ended specificity) for their assembly. The stiffer motif was provided by the DNA double-crossover (DX) molecule[10], analogous, once again, to the double Holliday-junction intermediate formed during meiosis (MDX, Fig. 2a). This stiff molecule contains two double helices connected to each other twice through crossover points. It is possible to program DX molecules to produce a variety of patterned two-dimensional arrays just by controlling their sticky ends[11–13] (Fig. 2b).

DNA nanomachines

In addition to objects and arrays, a number of DNA-based nanomechanical devices have been made. The first device consisted of two DX molecules connected by a shaft with a special sequence that could be converted from normal right-handed DNA (known as B-DNA) to an unusual left-handed conformation, known as Z-DNA[14]. The two DX molecules lie on one side of the shaft before conversion and on opposite sides after conversion, which leads to a rotation. The problem

with this device is that it is activated by a small molecule, $Co(NH_3)_6^{3+}$, and with all devices sharing the same stimulus, an ordered collection of DX molecules would not produce a diversity of responses.

This problem was solved by Bernard Yurke and colleagues, who developed a protocol for a sequence-control device that has a tweezers-like motion[15]. The principle behind the device is that a so-called 'set' strand containing a non-pairing extension hybridizes to a DNA-paired structural framework and sets a conformation; another strand that is complementary to the 'set' strand is then added, which binds to both the pairing and non-pairing portions, and removes it from the structure, leaving only the framework.

A robust rotary device was developed based on this principle[16] (Fig. 3), in which different set strands can enter and set the conformation to different structural end-states. In this way, the conformation of the DNA device can readily be flipped back and forth simply by adding different set strands followed by their complements. A variety of different devices can be controlled by a diverse group of set strands.

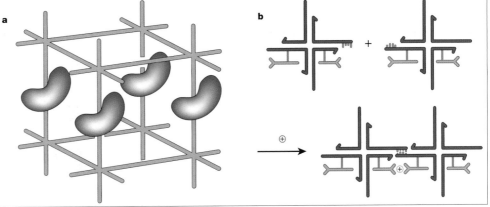

Figure 4 Applications of DNA scaffolds.
a, Scaffolding of biological macromolecules. A DNA box (red) is shown with protruding sticky ends that are used to organize boxes into crystals. Macromolecules are organized parallel to each other within the box, rendering them amenable to structure determination by X-ray crystallography.
b, DNA scaffolds to direct the assembly of nanoscale electrical circuits. Branched DNA junctions (blue) direct the assembly of attached nanoelectronic components (red), which are stabilized by the addition of a positively charged ion.

DNA as a scaffold

What is the purpose of constructing DNA arrays and nanodevices? One prominent goal is to use DNA as scaffolding to organize other molecules. For example, it may be possible to use self-assembled DNA lattices (crystals) as platforms to position biological macromolecules so as to study their structure by X-ray crystallography[4] (Fig. 4a). Towards this goal, programming of DNA has been used to bring protein molecules in proximity with each other to fuse multiple enzymatic activities[17]. However, the potential of this approach awaits the successful self-assembly of three-dimensional crystals.

Another goal is to use DNA crystals to assemble nanoelectronic components in two- or three-dimensional arrays[18] (Fig. 4b). DNA has been shown to organize metallic nanoparticles as a precursor to nanoelectronic assembly[19–22], but so far it has not been possible to produce multidimensional arrays containing nanoelectronic components with the high-structural order of the naked DNA arrays described earlier.

There has been some controversy over whether DNA can be used as an electrical conductor (for example, ref. 23), although the resolution of this debate is unlikely have any impact on the use of DNA as a scaffold. Recently, the effects of DNA conformational changes on conduction in the presence of an analyte were shown to have potential as a biosensor[24].

Replicating DNA components

A natural question to ask of any assembly system based on DNA is whether the components can be replicated. To produce branched DNA molecules whose branch points do not move, they must have different sequences in opposite branches but, as a consequence, these structures are not readily reproduced by DNA polymerase; the polymerase would produce complements to all strands present, leading only to double helical molecules. One option is to use topological tricks to convert structures like the DNA cube into a long single strand by adding extra stretches of DNA bases. The single strand could then be replicated by DNA polymerase and the final replicated product induced to fold into the original shape, with any extraneous segments cleaved using restriction enzymes. Although this would produce a molecule with sticky ends ready to participate in self-assembly, it would be a cumbersome process[25].

Günter von Kiedrowski and colleagues have recently developed a way of replicating short, simple DNA branches in a mixed organic–DNA species. Their branched molecule consists of three DNA single strands bonded to an organic triangle-shaped linker. To replicate the branched molecule, the single-stranded complement of each of these strands is bound to the molecule, so that one end of each complement molecule is close to the same end of the other complement molecule. In the final step, the juxtaposed complements are connected together by bonding their neighbouring ends to another molecule of the organic linker[26]. Extension of this system to the next level, such as objects like the cube, will need to solve topological problems involved in the separation of the two components, or it will be limited to unligated systems.

Future prospects

Many separate capabilities of DNA nanotechnology have been prototyped — it is now time to extend and integrate them into useful systems. Combining sequence-dependent devices with nanoscale arrays will provide a system with a vast number of distinct, programmable structural states, the *sine qua non* of nanorobotics. A key step in realizing these goals is to achieve highly ordered three-dimensional arrays, both periodic and, ultimately, algorithmic.

Interfacing with top-down nanotechnology will extend markedly the capabilities of the field. It also will be necessary to integrate biological macromolecules or other macromolecular complexes into DNA arrays in order to make practical systems with nanoscale components. Likewise, the inclusion of electronic components in highly ordered arrays will enable the organization of nanoelectronic circuits. Chemical function could be added to DNA arrays by adding nucleic acid species evolved *in vitro* to have specific binding properties ('aptamers') or enzymatic activities ('ribozymes' or 'DNAzymes'). A further area that has yet to have an impact on DNA nanotechnology is

Box 1
DNA computers

An assembly of DNA strands can process data in a similar way as an electronic computer, and has the potential to solve far more complex problems and store a greater amount of information, for substantially less energy costs than do electronic microprocessors. DNA-based computation dates from Leonard Adleman's landmark report in 1994 (ref. 27), where he used DNA to solve the 'Hamiltonian path' problem, a variant of the 'travelling salesman' problem. The idea is to establish whether there is a path between two cities, given an incomplete set of available roads. Adleman used strands of DNA to represent cities and roads, and encoded the sequences so that a strand representing a road would connect (according to the rules of base pairing) to any two strands representing a city. By mixing together the strands, joining the cities connected by roads, and weeding out any 'wrong answers', he showed that the strands could self-assemble to solve the problem.

It is impossible to separate DNA nanotechnology from DNA-based computation: many researchers work in both fields and the two communities have a symbiotic relationship. The first link between DNA computation and DNA nanotechnology was established by Erik Winfree, who suggested that short branched DNA molecules could be 'programmed' to undergo algorithmic self-assembly and thus serve as the basis of computation[28].

Periodic building blocks of matter, such as the DNA molecules shown in Fig. 1a, represent the simplest algorithm for assembly. All components are parallel, so what is on one side of a component is also on the other side, and in every direction. Given this parallelism, if the right side complements the left, the top complements the bottom and the front complements the back, a crystal should result. Even more complex algorithms are possible if one uses components of the same shape, but with different sticky ends. For example, Winfree has shown that, in principle, DNA tiles can be used to 'count' (see figure below) by creating borders with programmable sizes for one-, two- and possibly three-dimensional assemblies[29]. If this scheme can be realized, self-assembly of precisely sized nanoscale arrays will be possible. A computation using self-assembly has been prototyped in one dimension, thereby lending some credence to the viability of algorithmic assembly[30].

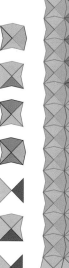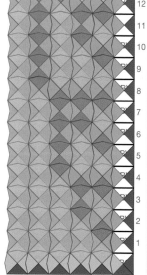

Box 1 Figure Counting with self-assembled DNA tiles. DNA tiles are represented by squares with coloured edges that are protruded or indented. Seven component tiles are shown on the left: three border tiles on the bottom and four tiles with the values **0** or **1**. The array illustrates binary counting from 1 (bottom row) to 12 (top row). Assembly is assumed to proceed by forming the reverse L-shaped border first, followed by binding the tiles that fit into the sites containing two (but not one) edges. Thus, the border determines the **1** tile in its bend, then that **1** tile and the horizontal-border tile on its left determine the **0** tile that fits, while the **1** tile and the vertical-border tile above it determine the (different) **0** tile that fits. (Adapted from ref. 29.)

combinatorial synthesis, which may well lead to greater diversity of integrated components. DNA-based computation and algorithmic assembly is another active area of research, and one that is impossible to separate from DNA nanotechnology (see Box 1).

The field of DNA nanotechnology has attracted an influx of researchers over the past few years. All of those involved in this area have benefited from the biotechnology enterprise that produces DNA-modifying enzymes and unusual components for synthetic DNA molecules. It is likely that applications in structural DNA nanotechnology ultimately will use variants on the theme of DNA (for example, peptide nucleic acids, containing an unconventional synthetic peptide backbone and nucleic acid bases for side chains), whose properties may be better suited to particular types of applications.

For the past half-century, DNA has been almost exclusively the province of biologists and biologically oriented physical scientists, who have studied its biological impact and molecular properties. During the next 50 years, it is likely they will be joined by materials scientists, nanotechnologists and computer engineers, who will exploit DNA's chemical properties in a non-biological context. □

doi:10.1038/nature01406

1. Hoffmann, R. DNA as clay. *Am. Sci.* **82**, 308–311 (1994).
2. Cuberes, M. T., Schlittler, R. R. & Gimzewski, J. K. Room-temperature repositioning of individual C-60 molecules at Cu steps: operation of a molecular counting device. *Appl. Phys. Lett.* **69**, 3016–3018 (1996).
3. Caruthers, M. H. Gene synthesis machines: DNA chemistry and its uses. *Science* **230**, 281–285 (1985).
4. Seeman, N. C. Nucleic acid junctions and lattices. *J. Theor. Biol.* **99**, 237–247 (1982).
5. Seeman, N. C. Molecular craftwork with DNA. *Chem. Intell.* **1**, 38–47 (1995).
6. Jaeger, L., Westhof, E. & Leontis, N. B. Tecto-RNA: modular assembly units for the construction of RNA nano-objects. *Nucleic Acids Res.* **29**, 455–463 (2001).
7. Zhang, X., Yan, H., Shen, Z. & Seeman, N. C. Paranemic cohesion of topologically-closed DNA molecules. *J. Am. Chem. Soc.* **124**, 12940–12941 (2002).
8. Chen, J. & Seeman, N. C. The synthesis from DNA of a molecule with the connectivity of a cube. *Nature* **350**, 631–633 (1991).
9. Seeman, N. C. Nucleic acid nanostructures and topology. *Angew. Chem. Int. Edn Engl.* **37**, 3220–3238 (1998).
10. Li, X., Yang, X., Qi, J. & Seeman, N. C. Antiparallel DNA double crossover molecules as components for nanoconstruction. *J. Am. Chem. Soc.* **118**, 6131–6140 (1996).
11. Winfree, E., Liu, F., Wenzler, L.A. & Seeman, N.C. Design and self-assembly of two-dimensional DNA crystals. *Nature* **394**, 539–544 (1998).
12. Mao, C., Sun, W. & Seeman, N. C. Designed two-dimensional DNA Holliday junction arrays visualized by atomic force microscopy. *J. Am. Chem. Soc.* **121**, 5437–5443 (1999).
13. LaBean, T. *et al.* The construction, analysis, ligation and self-assembly of DNA triple crossover complexes. *J. Am. Chem. Soc.* **122**, 1848–1860 (2000).
14. Mao, C., Sun, W., Shen, Z. & Seeman, N. C. A DNA nanomechanical device based on the B–Z transition. *Nature* **397**, 144–146 (1999).
15. Yurke, B., Turberfield, A. J., Mills, A. P. Jr, Simmel, F. C. & Newmann, J. L. A DNA-fuelled molecular machine made of DNA. *Nature* **406**, 605–608 (2000).
16. Yan, H., Zhang, X., Shen, Z. & Seeman, N. C. A robust DNA mechanical device controlled by hybridization topology. *Nature* **415**, 62–65 (2002).
17. Niemeyer, C. M., Koehler, J. & Wuerdemann, C. DNA-directed assembly of bi-enzymic complexes from *in vivo* biotinylated NADP(H):FMN oxidoreductase and luciferase. *ChemBioChem* **3**, 242–245 (2002).
18. Robinson, B. H. & Seeman, N. C. The design of a biochip: a self-assembling molecular-scale memory device. *Protein Eng.* **1**, 295–300 (1987).
19. Keren, K. *et al.* Sequence-specific molecular lithography on single DNA molecules. *Science* **297**, 72–75 (2002).
20. Alivisatos, A. P. *et al.* Organization of 'nanocrystal molecules' using DNA. *Nature* **382**, 609–611 (1996).
21. Taton, T. A., Mucic, R. C., Mirkin, C. A. & Letsinger, R. L. The DNA-mediated formation of supramolecular mono- and multilayered nanoparticle structures. *J. Am. Chem. Soc.* **122**, 6305–6306 (2000).
22. Pena, S. R. N., Raina, S., Goodrich, G. P., Fedoroff, N. V. & Keating, C. D. Hybridization and enzymatic extension of Au nanoparticle-bound oligonucleotides. *J. Am. Chem. Soc.* **124**, 7314–7323 (2002).
23. Dekker, C. & Ratner, M. A. Electronic properties of DNA. *Phys. World* **14**, 29–33 (2001).
24. Fahlman, R. P. & Sen, D. DNA conformational switches as sensitive electronic sensors of analytes. *J. Am. Chem. Soc.* **124**, 4610–4616 (2002).
25. Seeman, N. C. The construction of 3-D stick figures from branched DNA. *DNA Cell Biol.* **10**, 475–486 (1991).
26. Eckardt, L. H. *et al.* Chemical copying of connectivity. *Nature* **420**, 286 (2002).
27. Adleman, L. Molecular computation of solutions to combinatorial problems. *Science* **266**, 1021–1024 (1994).
28. Winfree, E. in *DNA Based Computers. Proceedings of a DIMACS Workshop, April 4, 1995, Princeton University* (eds Lipton, R. J & Baum, E. B.) 199–219 (American Mathematical Society, Providence, 1996).
29. Winfree, E. Algorithmic self-assembly of DNA: theoretical motivations and 2D assembly experiments. *J. Biol. Mol. Struct. Dynamics Conversat.* **11 2**, 263–270 (2000).
30. Mao, C., LaBean, T., Reif, J. H. & Seeman, N. C. Logical computation using algorithmic self-assembly of DNA triple crossover molecules. *Nature* **407**, 493–496 (2000).

Acknowledgements

This work has been supported by grants from the National Institute of General Medical Sciences, the Office of Naval Research, the National Science Foundation, and the Defense Advanced Research Projects Agency/Air Force Office of Scientific Research.

Original reference: Nature **421**, 427–431 (2003).

DNA replication and recombination

Bruce Alberts

National Academy of Sciences, 2101 Constitution Avenue, Washington DC 20418, USA

Knowledge of the structure of DNA enabled scientists to undertake the difficult task of deciphering the detailed molecular mechanisms of two dynamic processes that are central to life: the copying of the genetic information by DNA replication, and its reassortment and repair by DNA recombination. Despite dramatic advances towards this goal over the past five decades, many challenges remain for the next generation of molecular biologists.

"Though facts are inherently less satisfying than the intellectual conclusions drawn from them, their importance should never be questioned." James D. Watson, 2002.

DNA carries all of the genetic information for life. One enormously long DNA molecule forms each of the chromosomes of an organism, 23 of them in a human. The fundamental living unit is the single cell. A cell gives rise to many more cells through serial repetitions of a process known as cell division. Before each division, new copies must be made of each of the many molecules that form the cell, including the duplication of all DNA molecules. DNA replication is the name given to this duplication process, which enables an organism's genetic information — its genes — to be passed to the two daughter cells created when a cell divides. Only slightly less central to life is a process that requires dynamic DNA acrobatics, called homologous DNA recombination, which reshuffles the genes on chromosomes. In reactions closely linked to DNA replication, the recombination machinery also repairs damage that inevitably occurs to the long, fragile DNA molecules inside cells (see accompanying article by Friedberg, page 122).

The model for the DNA double helix[1] proposed by James Watson and Francis Crick is based on two paired DNA strands that are complementary in their nucleotide sequence. The model had striking implications for the processes of DNA replication and DNA recombination. Before 1953, there had been no meaningful way of even speculating about the molecular mechanisms of these two central genetic processes. But the proposal that each nucleotide in one strand of DNA was tightly base-paired with its complementary nucleotide on the opposite strand — either adenine (A) with thymine (T), or guanine (G) with cytosine (C) — meant that any part of the nucleotide sequence could act as a direct template for the corresponding portion of the other strand. As a result, any part of the sequence can be used either to create or to recognize its partner nucleotide sequence — the two functions that are central for DNA replication and DNA recombination, respectively.

In this review, I discuss how the discovery of the structure of DNA half a century ago opened new avenues for understanding the processes of DNA replication and recombination. I shall also emphasize how, as our understanding of complex biological molecules and their interactions increased over the years, there have been profound changes in the way that biologists view the chemistry of life.

Structural features of DNA

The research that immediately followed the discovery of the double helix focused primarily on understanding the structural properties

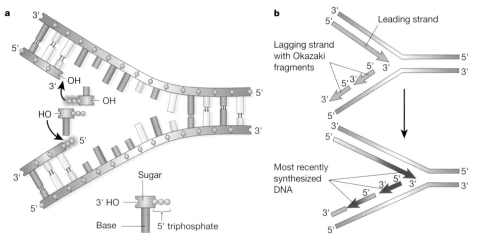

Figure 1 The DNA replication fork. **a**, Nucleoside triphosphates serve as a substrate for DNA polymerase, according to the mechanism shown on the top strand. Each nucleoside triphosphate is made up of three phosphates (represented here by yellow spheres), a deoxyribose sugar (beige rectangle) and one of four bases (differently coloured cylinders). The three phosphates are joined to each other by high-energy bonds, and the cleavage of these bonds during the polymerization reaction releases the free energy needed to drive the incorporation of each nucleotide into the growing DNA chain. The reaction shown on the bottom strand, which would cause DNA chain growth in the 3′ to 5′ chemical direction, does not occur in nature. **b**, DNA polymerases catalyse chain growth only in the 5′ to 3′ chemical direction, but both new daughter strands grow at the fork, so a dilemma of the 1960s was how the bottom strand in this diagram was synthesized. The asymmetric nature of the replication fork was recognized by the early 1970s: the 'leading strand' grows continuously, whereas the 'lagging strand' is synthesized by a DNA polymerase through the backstitching mechanism illustrated. Thus, both strands are produced by DNA synthesis in the 5′ to 3′ direction. (Redrawn from ref. 27, with permission.)

of the molecule. DNA specifies RNA through the process of gene transcription, and the RNA molecules in turn specify all of the proteins of a cell. This is the 'central dogma' of genetic information transfer[2]. Any read-out of genetic information — whether it be during DNA replication or gene transcription — requires access to the sequence of the bases buried in the interior of the double helix. DNA strand separation is therefore critical to DNA function. Thus, the Watson–Crick model drove scientists to a search for conditions that would disrupt the hydrogen bonds joining the complementary base pairs, so as to separate the two strands of the DNA double helix.

Physical chemists found that heating a solution of DNA to temperatures near boiling (100 °C), or subjecting it to extremes of pH, would cause the strands to separate — a change termed 'DNA denaturation'. The so-called 'melting temperature' (or T_m) of a stretch of DNA sequence depends on its nucleotide composition: those DNAs with a larger proportion of G–C base pairs exhibit a higher T_m because of the three hydrogen bonds that Watson and Crick had predicted to hold a G–C base pair together, compared with only two for the A–T base pair. At physiological salt concentrations, the T_m of mammalian DNA is nearly 90 °C, owing to the particular mix of its base pairs (47% G–C and 53% A–T)[3].

Initially it seemed inconceivable that, once separated from its complementary partner, a DNA strand could reform a double helix again. In a complex mixture of DNA molecules, such a feat would require finding the one sequence match amongst millions during random collisions with other sequences, and then rapidly rewinding with a new partner strand. The dramatic discovery of this unexpected phenomenon[4], called 'DNA renaturation', shed light on how sequences could be rearranged by DNA recombination. And it also provided a critical means by which DNA could be manipulated in the laboratory. The annealing of complementary nucleotide sequences, a process called hybridization, forms the basis of several DNA technologies that helped launch the biotechnology industry and modern genomics. These include gene cloning, genomic sequencing, and DNA copying by the polymerase chain reaction (see article by Hood and Galas on page 130).

The arrangement of DNA molecules in chromosomes presented another mystery for scientists: a long, thin molecule would be highly sensitive to shear-induced breakage, and it was hard to imagine that a mammalian chromosome might contain only a single DNA molecule. This would require that a typical chromosome be formed from a continuous DNA helix more than 100 million nucleotide pairs long

— a massive molecule weighing more than 100 billion daltons, with an end-to-end distance of more than 3 cm. How could such a giant molecule be protected from accidental fragmentation in a cell only microns in diameter, while keeping it organized for efficient gene readout and other genetic functions?

There was no precedent for such giant molecules outside the world of biology. But in the early 1960s, autoradiographic studies revealed that the chromosome of the bacterium *Escherichia coli* was in fact a single DNA molecule, more than 3 million nucleotide pairs in length[5]. And when — more than a decade later — innovative physical techniques demonstrated that a single huge DNA molecule formed the basis for each mammalian chromosome[6], the result was welcomed by scientists with little surprise.

DNA replication forks

How is the enormously long double-stranded DNA molecule that forms a chromosome accurately copied to produce a second identical chromosome each time a cell divides? The template model for DNA replication, proposed by Watson and Crick in 1953 (ref. 7), gained universal acceptance after two discoveries in the late 1950s. One was an elegant experiment using density-labelled bacterial DNAs that confirmed the predicted template–anti-template scheme[8]. The other was the discovery of an enzyme called DNA polymerase, which uses one strand of DNA as a template to synthesize a new complementary strand[9]. Four deoxyribonucleoside triphosphate nucleotides — dATP, dTTP, dGTP and dCTP — are the precursors to a new daughter DNA strand, each nucleotide selected by pairing with its complementary nucleotide (T, A, C or G, respectively) on the parental template strand. The DNA polymerase was shown to use these triphosphates to add nucleotides one at a time to the 3′ end of the newly synthesized DNA molecule, thereby catalysing DNA chain growth in the 5′ to 3′ chemical direction.

Although the synthesis of short stretches of DNA sequence on a single-stranded template could be demonstrated in a test tube, how an enormous, twisted double-stranded DNA molecule is replicated was a puzzle. Inside the cell, DNA replication was observed to occur at a Y-shaped structure, called a 'replication fork', which moves steadily along a parental DNA helix, spinning out two daughter DNA helices behind it (the two arms of the 'Y')[5]. As predicted by Watson and Crick, the two strands of the double helix run in opposite chemical directions. Therefore, as a replication fork moves, DNA polymerase can move continuously along only one arm of the Y — the arm on

Box 1
Core proteins at the DNA replication fork

Proteins at the Y-shaped DNA replication fork are illustrated schematically in panel **a** of the figure below, but in reality, the fork is folded in three dimensions, producing a structure resembling that of the diagram in the inset **b** (cartoons redrawn from ref. 27, with permission).

Focusing on the schematic illustration in **a**, two DNA polymerase molecules are active at the fork at any one time. One moves continuously to produce the new daughter DNA molecule on the leading strand, whereas the other produces a long series of short 'Okazaki DNA fragments' on the lagging strand. Both polymerases are anchored to their template by polymerase accessory proteins, in the form of a sliding clamp and a clamp loader.

A DNA helicase, powered by ATP hydrolysis, propels itself rapidly along one of the template DNA strands (here the lagging strand), forcing open the DNA helix ahead of the replication fork. The helicase exposes the bases of the DNA helix for the leading-strand polymerase to copy. DNA topoisomerase enzymes facilitate DNA helix unwinding.

In addition to the template, DNA polymerases need a pre-existing DNA or RNA chain end (a primer) onto which to add each nucleotide. For this reason, the lagging strand polymerase requires the action of a

DNA primase enzyme before it can start each Okazaki fragment. The primase produces a very short RNA molecule (an RNA primer) at the 5' end of each Okazaki fragment onto which the DNA polymerase adds nucleotides. Finally, the single-stranded regions of DNA at the fork are covered by multiple copies of a single-strand DNA-binding protein, which hold the DNA template strands open with their bases exposed.

In the folded fork structure shown in the inset, the lagging-strand DNA polymerase remains tied to the leading-strand DNA polymerase. This allows the lagging-strand polymerase to remain at the fork after it finishes the synthesis of each Okazaki fragment. As a result, this polymerase can be used over and over again to synthesize the large number of Okazaki fragments that are needed to produce a new DNA chain on the lagging strand.

In addition to the above group of core proteins, other proteins (not shown) are needed for DNA replication. These include a set of initiator proteins to begin each new replication fork at a replication origin, an RNAseH enzyme to remove the RNA primers from the Okazaki fragments, and a DNA ligase to seal the adjacent Okazaki fragments together to form a continuous DNA strand.

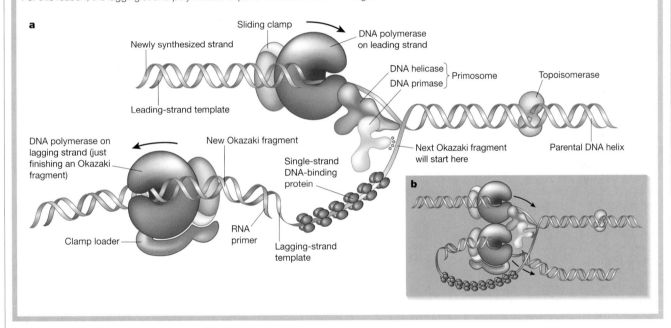

which the new daughter strand is being elongated in the 5' to 3' chemical direction. On the other arm, the new daughter strand would need to be produced in the opposite, 3' to 5' chemical direction (Fig. 1a). So, whereas Watson and Crick's central predictions were confirmed at the end of the first decade of research that followed their landmark discovery, the details of the DNA replication process remained a mystery.

Reconstructing replication

The mystery was solved over the course of the next two decades, a period in which the proteins that constitute the central players in the DNA replication process were identified. Scientists used a variety of experimental approaches to identify an ever-growing set of gene products thought to be critical for DNA replication. For example, mutant organisms were identified in which DNA replication was defective, and genetic techniques could then be used to identify specific sets of genes required for the replication process[10–12]. With the aid of the proteins specified by these genes, 'cell-free' systems were established, where the process was re-created *in vitro* using purified

components. Initially, proteins were tested in a 'partial replication reaction', where only a subset of the protein machinery required for the full replication process was present, and the DNA template was provided in a single-stranded form[13]. New proteins that were identified were added one at a time or in combination to test their effects on the catalytic activity of DNA polymerase. Further advances in understanding replication then depended on creating more complex *in vitro* systems, in which, through the addition of a larger set of purified proteins, double-stranded DNA could eventually be replicated[14–15].

Today, nearly every process inside cells — from DNA replication and recombination to membrane vesicle transport — is being studied in an *in vitro* system reconstructed from purified components. Although laborious to establish, such systems enable the precise control of both the concentration and the detailed structure of each component. Moreover, the 'noise' in the natural system caused by side reactions — because most molecules in a cell are engaged in more than one type of reaction — is avoided by eliminating the proteins that catalyse these other reactions. In essence, a small fraction of the cell can be re-created as a

Box 2
DNA recombination

Homologous DNA recombination involves an exchange between two DNA double helices that causes a section of each helix to be exchanged with a section of the other, as illustrated schematically in panel **a** in the figure below (redrawn from ref. 27, with permission). Critical to the reaction is the formation of a heteroduplex joint at the point where the two double helices have been broken and then joined together. To form this joint, which glues two previously separate molecules together, a strand from one helix must form base pairs with a complementary strand from the second helix. This requires that the two DNA helices that recombine have a very similar sequence of nucleotides, that is, they must be homologous.

The DNA double helix poses a major problem for the DNA recombination process, because the bases that need to pair to form a heteroduplex joint are buried in the interior of the helix. How can two DNA helices recognize that they are homologous, in order to begin a recombination event, if their bases are not exposed?

The breakthrough came from the isolation and characterization of the RecA protein[17] from the bacterium *Escherichia coli*, which would turn out to be the prototype for a family of strand-exchange proteins that is present in all organisms, from bacteria to humans. The human equivalent of the RecA protein is the Rad51 protein. These proteins catalyse the central synapsis step of homologous DNA recombination — the process that brings two matching DNA helices together and causes them to exchange parts, resulting in either the reassortment or the repair of genetic information (panel **b** below). Powered by the energy generated from ATP hydrolysis, the RecA protein assembles into long filaments on a single-strand DNA molecule (brown strand). Because the RecA protein has a second DNA-binding site that recognizes a DNA double helix, a RecA-coated strand has the remarkable ability to scan for a complementary strand in any double helix (blue strand) that it encounters. Once found, the complementary strand is pulled from the helix to form a new 'hybrid helix' with the RecA-coated single strand, thereby initiating the formation of the heteroduplex joint needed for recombination, as illustrated schematically in panel **b** (RecA protein not shown).

DNA recombination makes it possible for a damaged chromosome to repair itself by using a second copy of the same genetic information as a guide. It also causes the extensive breakage and reunion of chromosomes that occurs during the development of eggs and sperm, which greatly increases the genetic variation produced by sexual reproduction. Many of the atomic details of the RecA protein reaction are still uncertain, remaining as a future challenge for scientists.

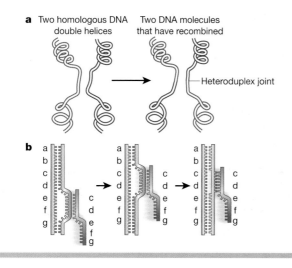

a Two homologous DNA double helices | Two DNA molecules that have recombined — Heteroduplex joint

b

bounded set of chemical reactions, making it fully amenable to precise study using all of the tools of physics and chemistry.

By 1980, multiprotein *in vitro* systems had enabled a detailed characterization of the replication machinery and solved the problem of how DNA is synthesized on both sides of the replication fork (Fig. 1b). One daughter DNA strand is synthesized continuously by a DNA polymerase molecule moving along the 'leading strand', while a second DNA polymerase molecule on the 'lagging strand' produces a long series of fragments (called Okazaki fragments)[16] which are joined together by the enzyme DNA ligase to produce a continuous DNA strand. As might be expected, there is a difference in the proteins required for leading- and lagging-strand DNA synthesis (see Box 1). Remarkably, the replication forks formed in these artificial systems could be shown to move at the same rapid rates as the forks inside cells (500 to 1,000 nucleotides per second), and the DNA template was copied with incredibly high fidelity[15].

As more and more proteins were found to function at the replication fork, comparisons could be made between the replication machinery of different organisms. Studies of the replication machinery in viruses, bacteria and eukaryotes revealed that a common set of protein activities drives the replication forks in each organism (Box 1). Each system consists of: a leading- and a lagging-strand DNA polymerase molecule; a DNA primase to produce the RNA primers that start each Okazaki fragment; single-strand DNA binding proteins that coat the template DNA and hold it in position; a DNA helicase that unwinds the double helix; and additional polymerase accessory proteins the tie the polymerases to each other and to the DNA template. As one progresses from a simple virus to more complex organisms, such as yeasts or mammals, the number of subunits that make up each type of protein activity tends to increase. For example, the total number of polypeptide subunits that form the core of the replication apparatus increases from four and seven in bacteriophages T7 and T4, respectively, to 13 in the bacterium *E. coli*. And it expands to at least 27 in the yeast *Saccharomyces cerevisiae* and in mammals. Thus, as organisms with larger genomes evolved, the replication machinery added new protein subunits, without any change in the basic mechanisms[15,18–20].

While the work I have described on DNA replication was advancing, other groups of researchers were establishing *in vitro* systems in which homologous DNA recombination could be reconstructed. The central player in these reactions was the RecA type of protein[17], named after the bacterial mutant defective in recombination that led to its discovery (Box 2).

Protein machines

As for all other aspects of cell biochemistry, the DNA replication apparatus has evolved over billions of years through 'trial and error'— that is, by random variation followed by natural selection. With time, one protein after another could be added to the mix of proteins active at the replication fork, presumably because the new protein increased the speed, control or accuracy of the overall replication process. In addition, the structure of each protein was fine-tuned by mutations that altered its amino acid sequence so as to increase its effectiveness. The end results of this unusual engineering process are the replication systems that we observe today in different organisms. The mechanism of DNA replication might therefore be expected to be highly dependent on random past events. But did evolution select for whatever works, with no need for elegance?

For the first 30 years after Watson and Crick's discovery, most researchers seemed to hold the view that cell processes could be sloppy. This view was encouraged by knowledge of the tremendous speed of movements at the molecular level (for example, it was known that a typical protein collides with a second molecule present at a concentration of 1 mM about 10^6 times per second). The rapid rates of molecular movement were thought initially to allow a process like DNA replication to occur without any organization of the proteins involved in three-dimensional space.

Quite to the contrary, molecular biologists now recognize that evolution has selected for highly ordered systems. Thus, for example, not only are the parts of the replication machinery held together in precise alignments to optimize their mutual interactions, but energy-driven changes in protein conformations are used to generate coordinated movements. This ensures that each of the successive steps in a complex process like DNA replication is closely coordinated with the next one. The result is an assembly that can be viewed as a 'protein machine'. For example, the DNA polymerase molecule on the lagging side of the replication fork remains bound to the leading-strand DNA polymerase molecule to ensure that the same lagging-strand polymerase is used over and over again for efficient synthesis of Okazaki fragments[18,20,21] (Box 1). And DNA replication is by no means unique. We now believe that nearly every biological process is catalysed by a set of ten or more spatially positioned, interacting proteins that undergo highly ordered movements in a machine-like assembly[22].

Protein machines generally form at specific sites in response to particular signals, and this is particularly true for protein machines that act on DNA. The replication, repair and recombination of the DNA double helix are often considered as separate, isolated processes. But inside the cell, the same DNA molecule is able to undergo any one of these reactions. Moreover, specific combinations of the three types of reactions occur. For instance, DNA recombination is often linked directly to either DNA replication or DNA repair[23]. For the integrity of a chromosome to be properly maintained, each specific reaction must be carefully directed and controlled. This requires that sets of proteins be assembled on the DNA and activated only where and when they are needed. Although much remains to be learned about how these choices are made, it seems that different types of DNA structures are recognized explicitly by specialized proteins that serve as 'assembly factors'. Each assembly factor then serves to nucleate a cooperative assembly of the set of proteins that forms a particular protein machine, as needed for catalysing a reaction appropriate to that time and place in the cell.

A view of the future

It has become customary, both in textbooks and in the regular scientific literature, to explain molecular mechanisms through simple two-dimensional drawings or 'cartoons'. Such drawings are useful for consolidating large amounts of data into a simple scheme, as illustrated in this review. But a whole generation of biologists may have become lulled into believing that the essence of a biological mechanism has been captured, and the entire problem therefore solved, once a researcher has deciphered enough of the puzzle to be able to draw a meaningful cartoon of this type.

In the past few years, it has become abundantly clear that much more will be demanded of scientists before we can claim to fully understand a process such as DNA replication or DNA recombination. Recent genome sequencing projects, protein-interaction mapping efforts and studies in cell signalling have revealed many more components and molecular interactions than were previously realized. For example, according to one recent analysis, S. cerevisiae, a single-celled 'simple' eukaryotic organism (which has about 6,000 genes compared with 30,000 in humans), uses 88 genes for its DNA replication and 49 genes for its DNA recombination[24].

To focus on DNA replication, fully understanding the mechanism will require returning to where the studies of DNA first began — in the realms of chemistry and physics. Detailed atomic structures of all relevant proteins and nucleic acids will be needed, and spectacular progress is being made by structural biologists, owing to increasingly powerful X-ray crystallography and nuclear magnetic resonance techniques. But the ability to reconstruct biological processes in a test tube with molecules whose precise structures are known is not enough. The replication process is both very rapid and incredibly accurate, achieving a final error rate of about one nucleotide in a billion. Understanding how the reactions between the many different proteins and other molecules are coordinated to create this result will require

that experimentalists determine all of the rate constants for the interactions between the various components, something that is rarely done by molecular biologists today. They can then use genetic engineering techniques to alter selected sets of these parameters, carefully monitoring the effect of these changes on the replication process.

Scientists will be able to claim that they truly understand a complex process such as DNA replication only when they can precisely predict the effect of changes in each of the various rate constants on the overall reaction. Because the range of experimental manipulations is enormous, we will need more powerful ways of deciding which such alterations are the most likely to increase our understanding. New approaches from the rapidly developing field of computational biology must therefore be developed — both to guide experimentation and to interpret the results.

The Watson–Crick model of DNA catalysed dramatic advances in our molecular understanding of biology. At the same time, its enormous success gave rise to the misleading view that many other complex aspects of biology might be similarly reduced to elegant simplicity through insightful theoretical analysis and model building. This view has been supplanted over subsequent decades, because most biological subsystems have turned out to be far too complex for their details to be predicted. We now know that nothing can substitute for rigorous experimental analyses. But traditional molecular and cell biology alone cannot bring a problem like DNA replication to closure. New types of approaches will be required, involving not only new computational tools, but also a greater integration of chemistry and physics[20,25]. For this reason, we urgently need to rethink the education that we are providing to the next generation of biological scientists[22,26]. □

doi:10.1038/nature01407

1. Watson, J. D. & Crick, F. H. C. A structure for deoxyribose nucleic acid. Nature 171, 737–738. (1953).
2. Crick, F. H. C. The biological replication of macromolecules. Symp. Soc. Exp. Biol. 12, 138–163 (1958).
3. Doty, P. Inside Nucleic Acids (Harvey Lecture, 1960) (Academic, New York, 1961).
4. Marmur, J. & Doty, P. Thermal renaturation of deoxyribonucleic acids. J. Mol. Biol. 3, 585–594 (1961).
5. Cairns, J. The bacterial chromosome and its manner of replication as seen by autoradiography. J. Mol. Biol. 6, 208–213 (1963).
6. Kavenoff, R., Klotz, L. C. & Zimm, B. H. On the nature of chromosome-sized DNA molecules. Cold Spring Harb. Symp. Quant. Biol. 38, 1–8 (1974).
7. Watson, J. D. & Crick, F. H. C. Genetical implications of the structure of deoxyribonucleic acid. Nature 171, 964–967 (1953).
8. Meselson, M. & Stahl, F. W. The replication of DNA in E. coli. Proc. Natl Acad. Sci. USA 44, 671–682 (1958).
9. Kornberg, A. Biological synthesis of DNA. Science 131, 1503-1508 (1960).
10. Epstein, R. H. et al. Physiological studies of conditional lethal mutants of bacteriophage T4D. Cold Spring Harb. Symp. Quant. Biol. 28, 375 (1963).
11. Bonhoeffer, F. & Schaller, H. A method for selective enrichment of mutants based on the high UV sensitivity of DNA containing 5-bromouracil. Biochem. Biophys. Res. Commun. 20, 93 (1965).
12. Kohiyama, M., Cousin, D., Ryter, A. & Jacob, F. Mutants thermosensibles d'Escherichia coli K/12. I. Isolement et caracterisation rapide. Ann. Inst. Pasteur 110, 465 (1966).
13. Huberman, J. A., Kornberg, A. & Alberts, B. M. Stimulation of T4 bacteriophage DNA polymerase by the protein product of T4 gene 32. J. Mol. Biol. 62, 39–52 (1971).
14. Morris, C. F., Sinha, N. K. & Alberts, B. M. Reconstruction of bacteriophage T4 DNA replication apparatus from purified components: rolling circle replication following de novo chain initiation on a single-stranded circular DNA template. Proc. Natl Acad. Sci. USA 72, 4800–4804 (1975).
15. Kornberg, A. & Baker, T. A. DNA Replication 2nd edn (Freeman, New York, 1992).
16. Okazaki R. et al. Mechanism of DNA chain growth: possible discontinuity and unusual secondary structure of newly synthesized chains. Proc. Natl Acad. Sci. USA 59, 598–605 (1968).
17. Radding, C. M. Recombination activities of E. coli RecA protein. Cell 25, 3–4 (1981).
18. Davey, M. J. & O'Donnell, M. Mechanisms of DNA replication. Curr. Opin. Chem. Biol. 4, 581–586 (2000).
19. Waga, S. & Stillman, B. The DNA replication fork in eukaryotic cells. Annu. Rev. Biochem. 67, 721–751 (1998).
20. Benkovic, S. J., Valentine, A. M. & Salinas F. Replisome-mediated DNA replication. Annu. Rev. Biochem. 70, 181–208 (2001).
21. Alberts, B. M. The DNA enzymology of protein machines. Cold Spring Harb. Symp. Quant. Biol. 49, 1–12 (1984).
22. Alberts, B. The cell as a collection of protein machines: preparing the next generation of molecular biologists. Cell 92, 291–294 (1998).
23. Radding, C. Colloquium introduction. Links between recombination and replication: vital roles of recombination. Proc. Natl Acad. Sci. USA 98, 8172 (2001).
24. Dwight, S. S. et al. Saccharomyces Genome Database (SGD) provides secondary gene annotation using the Gene Ontology (GO). Nucleic Acids Res. 30, 69–72 (2002).
25. Trakselis, M. A. & Benkovic, S. J. Intricacies in ATP-dependent clamp loading: variations across replication systems. Structure 9, 999–1004 (2001).
26. National Research Council. Bio2010: Undergraduate Education to Prepare Biomedical Research Scientists (The National Academies Press, Washington DC, 2002).
27. Alberts, B. et al. Molecular Biology of the Cell 4th edn (Garland, New York, 2002).

Original reference: Nature 421, 431–435 (2003).

DNA damage and repair

Errol C. Friedberg

Laboratory of Molecular Pathology, Department of Pathology, University of Texas Southwestern Medical Center, Dallas, Texas 75390-9072, USA (e-mail: friedberg.errol@pathology.swmed.edu)

The aesthetic appeal of the DNA double helix initially hindered notions of DNA mutation and repair, which would necessarily interfere with its pristine state. But it has since been recognized that DNA is subject to continuous damage and the cell has an arsenal of ways of responding to such injury. Although mutations or deficiencies in repair can have catastrophic consequences, causing a range of human diseases, mutations are nonetheless fundamental to life and evolution.

"We totally missed the possible role of … [DNA] repair although … I later came to realise that DNA is so precious that probably many distinct repair mechanisms would exist." Francis Crick, writing in *Nature*, 26 April 1974 (ref. 1).

This retrospective reflection by Francis Crick, penned two decades after he and James Watson reported the structure of DNA, hints at the early perception of DNA as a highly stable macromolecular entity. This prevailing view at the time significantly delayed serious consideration of biochemical processes such as mutation and repair. It was once suggested by Frank Stahl that "the possibility that … genes were … subject to the hurly-burly of both insult and clumsy efforts to reverse the insults, was unthinkable."[2]

But subsequent work on three 'R's of DNA metabolism — replication (copying of DNA prior to each cell division), recombination (exchanges between different DNA molecules in a cell) and repair (restoration of altered DNA to its normal state) — revealed the dynamic state of DNA. It became apparent that DNA in all living organisms continually incurs a myriad of types of damage, and that cells have devised ingenious mechanisms for tolerating and repairing the damage. Failure of these mechanisms can lead to serious disease consequences, as well illustrated in the human hereditary diseases xeroderma pigmentosum (XP), hereditary non-polyposis colon cancer (HNPCC) and some forms of breast cancer. XP is characterized by about a 10,000-fold increased risk of skin cancer associated with sunlight exposure; individuals with HNPCC manifest an increased hereditary predisposition to colon (and other) cancer.

The roots of repair

The early work on DNA damage and repair in the 1930s was stimulated by a small but prominent group of physicists[3]. As recounted by the geneticist Guido Pontecorvo, "in the years immediately preceding World War II something quite new happened: the introduction of ideas (not techniques) from the realm of physics into the realm of genetics, particularly as applied to the problems of size, mutability, and self-replication of genes"[4]. Seminal to this coalition between physics and biology in pre-war Germany was the collaboration between German physicists Karl Zimmer and Max Delbrück and the Russian geneticist Nikolai Timoféeff-Ressovsky[5]. Their partnership was stimulated by the work of Hermann Muller, a geneticist working

on the fruitfly *Drosophila* who first demonstrated that external agents, such as ionizing radiation, can cause mutations in living organisms[6].

Timoféeff-Ressovsky and Zimmer were interested primarily in how such small amounts of energy in the form of ionizing radiation (formally equivalent to no more than the amount of energy absorbed as heat by drinking a cup of hot tea) could have such profound biological effects[3]. Delbrück and Muller, on the other hand, were intrigued by whether such mutations could reveal insight into the physical nature of the gene.

In retrospect, it was inevitable that the deployment of physical (and later chemical) tools, such as ionizing and ultraviolet (UV) radiation, to study genes would in due course lead to questions as to how these agents damaged DNA[3]. And, once it was recognized that these interactions promoted deleterious effects on the structure and function of genes, to questions concerning how cells cope with damaged DNA. Zimmer wrote, "one cannot use radiations for elucidating the normal state of affairs without considering the mechanisms of their actions, nor can one find out much about radiation induced changes without being interested in the normal state of the material under investigation."[7]

Hints of the ability of living cells to recover from the lethal effect of UV radiation emerged as early as the mid-1930s[8]. But the discovery of a DNA-repair mechanism had to wait until the end of the 1940s, through the independent, serendipitous observations of Albert Kelner[9] working in Milislav Demerec's group at the Cold Spring Harbor Laboratory, and Renato Dulbecco[10] in Salvador Luria's laboratory at The University of Indiana. Neither Kelner nor Dulbecco set out to study damage to DNA or its repair. They were both using UV radiation as an experimental tool, but observed anomalous survival rates when cells or bacteriophage (bacteria-infecting viruses) were inadvertently exposed to long-wavelength light, either as sunlight or fluorescent light in their respective laboratories[9,10]. Their efforts to explain these confounding observations led to the discovery of the phenomenon now known as photoreacti-

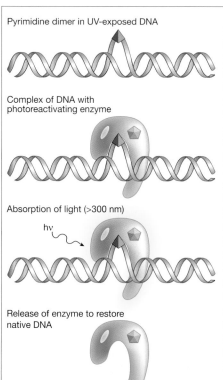

Pyrimidine dimer in UV-exposed DNA

Complex of DNA with photoreactivating enzyme

Absorption of light (>300 nm)

hv

Release of enzyme to restore native DNA

Figure 1
Photoreactivation reverses DNA damage. DNA exposed to ultraviolet (UV) radiation results in covalent dimerization of adjacent pyrimidines, typically thymine residues (thymine dimers), illustrated here as a purple triangle. These lesions are recognized by a photoreactivating enzyme, which absorbs light at wavelengths >300nm (such as fluorescent light or sunlight) and facilitates a series of photochemical reactions that monomerize the dimerized pyrimidines, restoring them to their native conformation.

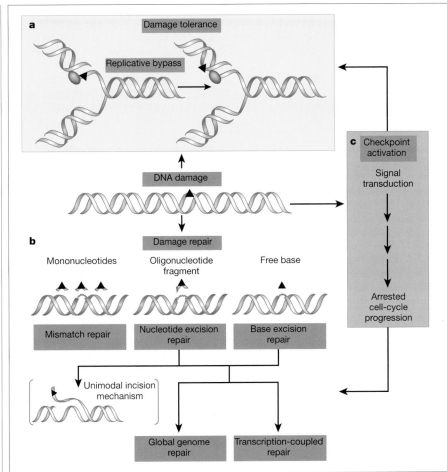

Figure 2 Responses to DNA damage. DNA damage (illustrated as a black triangle) results in either repair or tolerance. **a**, During damage tolerance, damaged sites are recognized by the replication machinery before they can be repaired, resulting in an arrest that can be relieved by replicative bypass (translesion DNA synthesis) (see Fig. 3). **b**, DNA repair involves the excision of bases and DNA synthesis (red wavy lines), which requires double-stranded DNA. Mispaired bases, usually generated by mistakes during DNA replication, are excised as single nucleotides during mismatch repair. A damaged base is excised as a single free base (base excision repair) or as an oligonucleotide fragment (nucleotide excision repair). Such fragments are generated by incisions flanking either side of the damaged base. Nucleotide excision repair can also transpire in some organisms by a distinct biochemical mechanism involving only a single incision next to a site of damage (unimodal incision). **c**, The cell has a network of complex signalling pathways that arrest the cell cycle and may ultimately lead to programmed cell death.

vation, whereby the DNA damage incurred by exposure to UV light is repaired by a light-dependent enzyme reaction[11] (Fig. 1).

Curiously, even with the elucidation of the structure of DNA only four years away, neither Dulbecco nor Watson — who was a graduate student in Luria's laboratory when Dulbecco stumbled on photoreactivation, and had himself examined the effects of ionizing radiation for his doctoral thesis[2] — thought about DNA repair. However, shortly after Watson and Crick reported on the DNA double helical structure, they noted the implications of the base-pairing rules for mutagenesis, stating "spontaneous mutation may be due to a base occasionally occurring in one of its less likely tautomeric forms"[12].

Tautomerism is the property of a compound that allows it to exist in two interconvertible chemical states; in the case of DNA bases, as either keto or enol forms. Watson and Crick had initially overlooked the complications of tautomerism and were trying unsuccessfully to construct their DNA model with the rare enol form of bases. It was only after Jerry Donohue, a former graduate student of Linus Pauling, pointed out to Watson that he should be using the more common keto form that the problem of how bases could stably pair was solved[13].

But no consideration was then given to the fact that the chemical lability of DNA implicit in tautomerism might have wider implications for the stability of genes. Indeed, the field gave little thought to the precise nature of DNA damage and its possible biological consequences. One must recall, however, that even at the time the DNA double helix was unveiled, its 'pathology' and the biological consequences thereof were far less compelling problems than deciphering the genetic code or understanding the essential features of DNA replication. Even mutagenesis — put to extensive use as a tool for determining the function of genes and their polypeptide products, and for defining the genetic code — was not widely considered in mechanistic terms until much later[11]. This is despite the fact that

the repair phenomenon of photoreactivation was known before the discovery of the structure of DNA.

A DNA duplex for redundancy

Watson and Crick noted, with infamous prophetic understatement, "it has not escaped our notice that the specific [base] pairing we have postulated immediately suggests a possible copying mechanism for the genetic material"[14]. However, it was not intuitively obvious that a double-stranded molecule should be required for DNA replication. In principle, a single-stranded chain could just as easily do. But the significance of the duplex DNA structure soon became apparent. It was shown that DNA replicates in a semi-conservative fashion, whereby each strand of the double helix pairs with a new strand generated by replication. This enables errors introduced during DNA replication to be corrected by a mechanism known as excision repair, which relies on the redundancy inherent in having two complementary strands of the genetic code. If the nucleotides on one strand are damaged, they can be excised and the intact opposite strand used as a template to direct repair synthesis of DNA[15] (Fig. 2).

Many paths to mutation and repair

The elucidation of the DNA structure provided the essential foundation for defining the different types of mutations arising from both spontaneous and environmental DNA damage that affect all living cells[12]. Once again, the insights of physicists featured prominently[3], including among others, Richard Setlow who identified thymine dimers as stable and naturally occurring DNA lesions arising in cells exposed to sunlight (UV radiation). Such lesions comprise a covalent joining of two adjacent thymine residues in the same DNA chain. They generate considerable distortion of the normal structure of DNA and seriously impede DNA transactions such as replication and transcription. The repair of these lesions could be monitored experimentally, and promoted the discovery by Setlow[16] and others[2] of excision repair in bacteria and higher organisms[17].

As the profusion of alterations in DNA became more widely recognized, scientists came to appreciate that the identification of any new type of naturally occurring base damage would, if one searched diligently enough, almost certainly lead to the discovery of one or more mechanisms for its repair or tolerance[2,18]. Such has indeed been the case. DNA repair now embraces not only the direct reversal of some types of damage (such as the enzymatic photoreactivation of thymine dimers), but also multiple distinct mechanisms for excising damaged bases, termed nucleotide excision repair (NER), base excision repair (BER) and mismatch repair (MMR)[11] (Fig 2). The principle of all three mechanisms of repair involves splicing out the damaged region and inserting new bases to fill the gap, followed by ligation of the pieces.

The process of NER is biochemically complicated, involving as many as 30 distinct proteins in human cells that function as a large complex called the nucleotide excision repairosome. This 'repair machine' facilitates the excision of damaged nucleotides by generating bimodal incisions in the flanking regions and removing a fragment about 30 nucleotides in length[11] (Fig. 2). Damaged bases that are not recognized by the NER machinery are corrected by BER, whereby the bases are excised from the genome as free bases by a different set of repair enzymes. In MMR, incorrect bases incorporated as a result of mistakes during DNA replication are excised as single nucleotides by yet a third group of repair proteins (Fig. 2). Both NER and BER transpire by somewhat different mechanisms depending on whether the DNA damage is located in regions of the genome that are undergoing active gene expression (transcription-coupled repair) or are transcriptionally silent (global genome repair)[11,19].

In addition to the various modes of excision repair that evolved to cope with damaged bases or mistakes during replication, cells frequently suffer breakage of one or both chains of the DNA duplex[11]. Naturally occurring reactive oxygen molecules and ionizing radiation are prevalent sources of such damage[11]. Strand breaks must be repaired in order to maintain genomic integrity. In particular, double-strand breaks (DSBs) sever the chromosomes and are lethal unless repaired[11].

Several mechanisms for the repair of DSBs have been elucidated (Fig. 3). One of these involves swapping equivalent regions of DNA between homologous chromosomes — a process called recombination[11]. This type of exchange occurs naturally during meiosis, the special type of cell division that generates the germ cells (sperm and ova). It can also be used to repair a damaged site on a DNA strand by using information located on the undamaged homologous chromosome. This process requires an extensive region of sequence homology between the damaged and template strands. Multiple proteins are required for DSB repair by recombination and deficiencies in this repair mechanism can cause cancer. For example, mutation of at least one of these repair proteins (called BRCA1) causes hereditary breast cancer. An alternative mechanism for the repair of DSBs, called non-homologous end joining, also requires a multi-protein complex, and essentially joins broken chromosome ends in a manner that does not depend on sequence homology and may not be error free (Fig. 3).

Damage tolerance

Although insights into DNA repair have progressed at an impressive pace, especially in the past decade, an understanding of the mechanisms of mutagenesis — a phenomenon that, as mentioned earlier, was demonstrated experimentally before discovery of the structure of DNA — has lagged. A breakthrough came from the experimental demonstration that some mutations arise as a consequence of a cell's efforts to tolerate damage. In this situation, the base damage and/or strand breaks in DNA persist in the genome, but their potential for interfering with DNA replication and transcription is somehow mitigated.

One such damage-tolerance mechanism, called translesion DNA synthesis, involves the replication machinery bypassing sites of base damage, allowing normal DNA replication and gene expression to proceed downstream of the (unrepaired) damage[20] (Fig. 4). It involves specialized low-fidelity ('sloppy') DNA polymerases that are able to bypass DNA lesions that typically stall the high-fidelity polymerases required for DNA replication. To overcome the block, these 'sloppy copiers' add nucleotides to the replicating strand opposing the DNA lesion, thus allowing replication to continue, but nevertheless introducing mutations into the newly synthesized sequence[20].

Cell suicide

Recent years have witnessed the recognition that biological responsiveness to genetic insult embraces more than the repair and

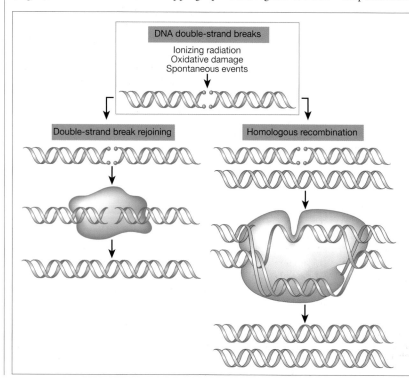

Figure 3 The repair of double-strand breaks in DNA. Double-strand breaks can result from exposure to ionizing radiation, oxidative damage and the spontaneous cleavage of the sugar-phosphate backbone of the DNA molecule. Their repair can be effected by either rejoining the broken ends (left) or by homologous recombination with a sister DNA molecule (right). Both processes involve different multi-protein complexes.

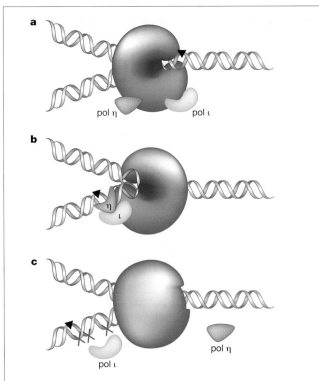

Figure 4 'Sloppy copiers' overcome blocks in replication caused by a DNA lesion (a process called translesion synthesis). **a**, The DNA replicative machinery (blue) stalls immediately behind a site of base damage (black triangle). Two specialized 'sloppy copier' polymerases (polη and polι) bind to the arrested replication complex. **b**, This interaction promotes a conformational change in the arrested replication machinery, placing polη in direct proximity to the site of base damage where it synthesizes across the lesion. **c**, Polη may then dissociate and allow polι to complete the process of replicative bypass by incorporating several more nucleotides (red crosses). Once the lesion has been completely bypassed, the replication machinery resumes DNA replication. As a result of this process, mutations to the DNA sequence are now incorporated into one strand.

tolerance of DNA damage. The exposure of cells to many DNA-damaging agents results in the transcriptional upregulation of a large number of genes, the precise function(s) of many of which remains to be established. Additionally, cells have evolved complex signalling pathways to arrest the progression of the cell cycle in the presence of DNA damage, thereby providing increased time for repair and tolerance mechanisms to operate[21] (Fig. 2c). Finally, when the burden of genomic insult is simply too large to be effectively met by the various responses discussed, cells are able to initiate programmed cell death (apoptosis), thereby eliminating themselves from a population that otherwise might suffer serious pathological consequences[22].

DNA damage and cancer

The 'somatic mutation hypothesis' of cancer embraces the notion that neoplastic transformation arises from mutations that alter the function of specific genes (now called oncogenes and tumour-suppressor genes) that are critical for cell division. This theory has its roots in correlations between chromosomal abnormalities and cancer first observed by the developmental biologist Theodore Boveri[23], who at the beginning of the twentieth century reported abnormal numbers of chromosomes (aneuploidy) in cancerous somatic cells.

The discovery of the structure of DNA progressed our understanding of tumorigenesis at several levels. Watson and Crick predicted from their DNA model that complementary base pairing had implications for recombination (the exchange of genetic

material between chromosome pairs): "the pairing between homologous chromosomes at meiosis may depend on pairing between specific bases". The genetic basis of many cancers is now known to arise from abnormal recombination events, such as chromosomal translocations, where a region of one chromosome is juxtaposed to another chromosome. Watson himself developed an early and ardent interest in cancer biology when he recognized that the experimentally tractable genomes of oncogenic viruses could provide important insights into the pathogenesis of cancer. Mutagenesis is now documented as a fundamental cornerstone of the molecular basis of all forms of cancer[24].

Arguably the most definitive validation of the somatic mutation hypothesis derives from the discovery that defective responses to DNA damage and the accumulation of mutations underlies two distinct types of hereditary cancer; skin cancer associated with defective NER and colorectal cancer associated with defective MMR[11]. In both instances, credit belongs to scholars of DNA repair.

In the late 1960s, James Cleaver providentially noted an article in the *San Francisco Chronicle* that reported the extreme proneness to skin cancer in individuals with XP, a rare sun-sensitive hereditary disease[2]. Cleaver was then searching for mammalian cell lines that were defective in excision repair, and his intuitive notion that XP individuals might be sunlight-sensitive and prone to cancer because they were genetically defective in excision repair proved to be correct[25].

The subsequent elucidation of the genes defective in XP patients[26], and their role in NER of damaged bases in human cells[11,27,28], represents a triumph of modern genetics and its application to molecular biology. The additional discovery that the process of NER in eukaryotes requires elements of the basic transcription apparatus[11] has yielded insights into the complex relationships between deficient DNA repair, defective transcription and hereditary human diseases[11].

A fascinating denouement to the skin-cancer predisposition in XP patients derives from the recent solution of the 'XP variant problem'. A significant fraction of XP individuals who are clinically indistinguishable from those defective in NER were found to be proficient in this repair process[11]. It was shown that DNA polymerase-η (polη), one of the specialized DNA polymerases capable of overcoming replication blocks at DNA lesions, is mutated in all XP-variant patients so far examined[29]. Not only does polη replicate past thymine dimers in DNA, but — unlike the other specialized DNA polymerases — it also correctly inserts adenine residues[29], thereby preventing mutations at sites of thymine dimers. Therefore, even in XP patients with functional NER, in the absence of polη one or more other bypass polymerases attempts to cope with arrested replication at thymine dimers, but does so inaccurately[29]. Thus, cancer predisposition in XP essentially derives from an excessive mutational burden in skin cells associated with exposure to sunlight. These mutations accumulate either because thymine dimers are not excised (owing to defective NER) or because in the absence of polη, dimers are inaccurately bypassed by other DNA polymerases[29].

The association between HNPCC and defective MMR was determined more-or-less simultaneously by a number of investigators. Paul Modrich[2] surmised that the instability of repeated sequences in DNA associated with defective MMR in bacteria[30] might be causally related to the DNA sequence instability observed in patients with HNPCC[31]. This led to the formal demonstration of defective MMR in this human hereditary disease and formed another persuasive validation of the somatic mutation theory of cancer[32–34].

A look to the future

The study of biological responsiveness to DNA damage embraces DNA repair, mutagenesis, damage tolerance, cell-cycle checkpoint control, programmed cell death, and other cellular responses to genomic insult. This integrated field is now deciphering the complex regulatory pathways transduced by signalling mechanisms that detect DNA damage and/or arrested DNA replication. As these pathways become better understood, parallel technological gains in gene therapy and therapeutic intervention by rational drug design will offer new strategies for blocking the unwanted consequences of DNA damage, especially cancer.

We must remember, however, that while evolution could not have transpired without robust cellular mechanisms to ameliorate the most serious consequences of spontaneous and environmental DNA damage, the process of evolution mandates that the genetic diversification on which Darwinian selection operates be maintained constantly. Thus, life is necessarily a delicate balance between genomic stability and instability—and of mutation and repair. □

doi:10.1038/nature01408

1. Crick, F. The double helix: a personal view. *Nature* **248**, 766–769 (1974).
2. Friedberg, E. C. *Correcting the Blueprint of Life: An Historical Account of the Discovery of DNA Repair Mechanisms* (Cold Spring Harbor Laboratory Press, Cold Spring Harbor, NY, 1997).
3. Friedberg, E. C. The intersection between the birth of molecular biology and the DNA repair and mutagenesis field. *DNA Repair* **1**, 855–867 (2002).
4. Pontecorvo, G. *Trends in Genetic Analysis* (Columbia Univ. Press, New York, 1958).
5. Timoféeff-Ressovsky, N. W., Zimmer, K. G. & Delbrück, M. Über die Natur der Genmutation und der Genkostruktur. *Nachr. Ges. Wiss. Gottingen FG VI Biol. N.F.* **1**, 189–245 (1935).
6. Muller, H. J. Artificial transmutation of the gene. *Science* **66**, 84–87 (1927).
7. Zimmer, K. G. in *Phage and the Origins of Molecular Biology* Expanded edn (eds Cairns, J., Stent, G. S. & Watson, J. D.) (Cold Spring Harbor Laboratory Press, Cold Spring Harbor, NY, 1992).
8. Hollaender, A. & Curtis, J. T. Effect of sublethal doses of monochromatic ultraviolet radiation on bacteria in liquid suspension. *Proc. Soc. Exp. Biol. Med.* **33**, 61–62 (1935–36).
9. Kelner, A. Effect of visible light on the recovery of *Streptomyces griseus conidia* from ultraviolet irradiation injury. *Proc. Natl Acad. Sci. USA* **35**, 73–79 (1949).
10. Dulbecco, R. Reactivation of ultra-violet-inactivated bacteriophage by visible light. *Nature* **162**, 949–950 (1949).
11. Friedberg, E. C., Walker, G. C. & Siede, W. *DNA Repair and Mutagenesis* (American Society of Microbiology Press, Washington DC, 1995).
12. Watson, J. D. & Crick, F. H. C. Genetical implications of the structure of deoxyribonucleic acid. *Nature* **171**, 964–967 (1953).
13. Watson, J. D. *The Double Helix. A Personal Account of the Discovery of the Structure of DNA* (Atheneum, New York, 1968).
14. Watson, J. D. & Crick, F. H. C. A structure for deoxyribose nucleic acid. *Nature* **171**, 737–738 (1953).
15. Hanawalt, P. C. & Haynes, R. H. The repair of DNA. *Sci. Am.* **216**, 36–43 (1967).
16. Setlow, R. B. & Carrier, W. L. The disappearance of thymine dimers from DNA: an error-correcting mechanism. *Proc. Natl Acad. Sci. USA* **51**, 226–231 (1964).
17. Rasmussen, D. E. & Painter, R. B. Evidence for repair of ultra-violet damaged deoxyribonucleic acid in cultured mammalian cells. *Nature* **203**, 1360–1362 (1964).
18. Lindahl, T. Instability and decay of the primary structure of DNA. *Nature* **362**, 709–715 (1993).
19. Le Page, F. *et al.* Transcription-coupled repair of 8-oxoguanine: requirement for XPG, TFIIH, and CSB and implications for Cockayne syndrome. *Cell* **101**, 159–171 (2000).
20. Goodman, M. F. Error-prone repair DNA polymerases in prokaryotes and eukaryotes. *Annu. Rev. Biochem.* **71**, 17–50 (2002).
21. Zhou, B.-B. S. & Elledge, S. J. The DNA damage response: putting checkpoints in perspective. *Nature* **408**, 433–439 (2000).
22. Cory, S. & Adams J. M. The Bcl2 family: regulators of the cellular life-or-death switch. *Nature Rev. Cancer* **2**, 647–656 (2002).
23. Boveri, T. *The Origin of Malignant Tumors* (Williams and Wilkins, Baltimore, MD, 1929).
24. Vogelstein, B. & Kinzler, K. W. (eds) *The Genetic Basis of Human Cancer* (McGraw-Hill, New York, 1998).
25. Cleaver, J. E. Defective repair replication of DNA in xeroderma pigmentosum. *Nature* **218**, 652–656 (1968).
26. De Weerd-Kastelein, E. A., Keijzer, W. & Bootsma, D. Genetic heterogeneity of xeroderma pigmentosum demonstrated by somatic cell hybridization. *Nature New Biol.* **238**, 80–83 (1972).
27. Wood, R. D. DNA repair in eukaryotes. *Annu. Rev. Biochem.* **65**, 135–167 (1996).
28. Sancar, A. DNA excision repair. *Annu. Rev. Biochem.* **65**, 43–81 (1996).
29. Masutani, C. *et al.* Xeroderma pigmentosum variant: from a human genetic disorder to a novel DNA polymerase. *Cold Spring Harbor Symp. Quant. Biol.* **65**, 71–80 (2000).
30. Levinson, G., & Gutman, G. A. High frequencies of short frameshifts in poly-CA/TG tandem repeats borne by bacteriophage M13 in *Escherichia coli* K-12. *Nucleic Acids Res.* **15**, 5323–5338 (1987).
31. Aaltonen, L. A. *et al.* Clues to the pathogenesis of familial colorectal cancer. *Science* **260**, 812–816 (1993).
32. Fishel, R. *et al.* The human mutator gene homolog MSH2 and its association with hereditary nonpolyposis colon cancer. *Cell* **75**, 1027–1038 (1993).
33. Leach, F. S. *et al.* Mutations of a mutS homolog in hereditary nonpolyposis colorectal cancer. *Cell* **75**, 1215–1225 (1993).
34. Parsons, R. *et al.* Hypermutability and mismatch repair deficiency in RER+ tumor cells, *Cell* **75**, 1227–1236 (1993).

Acknowledgements
I thank numerous individuals for their constructive comments. This article is dedicated to James Watson with thanks for his inspiration as a writer and historian.

Original reference: *Nature* **421**, 436–440 (2003).

The double helix and immunology

Gustav J. V. Nossal

Department of Pathology, The University of Melbourne, Victoria 3010, Australia

The immune system can recognize and produce antibodies to virtually any molecule in the Universe. This enormous diversity arises from the ingenious reshuffling of DNA sequences encoding components of the immune system. Immunology is an example of a field completely transformed during the past 50 years by the discovery of the structure of DNA and the emergence of DNA technologies that followed.

"This short history of research in one area, lymphocyte receptors, is yet another witness to the power of DNA technology, and to the ability of this approach not only to explain known biological phenomena, but also to contribute to the discovery of new biological systems."
Susumu Tonegawa, Nobel lecture, 8 December 1987.

The double helix is all about biological information: how it is encoded, stored, replicated and used when required. Immunology, too, is about information. What genetic processes control the vast array of synthetic potential within an immune system capable of reacting specifically to virtually any microbe or foreign molecule? The secret lies in unique DNA processing that occurs during the development of lymphocyte cells, which are responsible for the specific immune response to a foreign agent (antigen). B lymphocytes produce antibodies (on their surface as well as secreted) and T lymphocytes mount cellular attacks on pathogenic infiltrators.

As lymphocytes develop, an array of short genes are rearranged and assembled together at the DNA level to form genes whose products recognize distinct antigens. As the process is mostly random, each lymphocyte makes different choices and thus the result is a vast repertoire of lymphocytes reactive to different antigens. This process has implications for antibody formation, cell-mediated immunity and malignancies of the immune system.

One B cell produces one antibody
At the beginning of the last century, Paul Ehrlich[1] recognized that the specificity of antibodies lay in the complementarity of their shapes to the antigen(s) on the microbe being recognized. He saw antibodies as cellular 'side chains', which budded out from the cell surface as what today we would term receptors. Karl Landsteiner[2] then demonstrated the exquisite specificity of antibodies, showing that animals could make antibodies to almost anything, including small synthetic organic molecules that had never previously existed in nature. Moreover, tiny structural changes in the antigen could lead to the production of a different antibody. It beggared belief that there could be so many different side chains. When antibodies were shown to be proteins, it seemed natural to conclude that a specific antibody molecule was shaped in close proximity to an antigen molecule much as plastic or sheet metal is moulded against a template. This 'direct template' hypothesis[3] held sway for several decades.

In 1955, Niels Jerne[4] published his natural selection theory of antibody formation, which postulated the random synthesis of a million or more different sorts of antibodies. When an antigen enters the body, it unites with an antibody that just happens to fit it, the antigen–antibody complex is taken up by a cell and the antibody

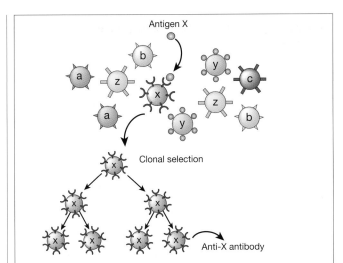

Figure 1 The clonal selection theory of antibody formation. Each B lymphocyte produces only one type of antibody receptor on its surface. An antigen recognizes one B lymphocyte out of a large repertoire. This triggers the rapid division and differentiation of the B cell to become a 'plasma cell', producing and secreting antibodies specific to the original antigen.

somehow acts as the template for the formation of more of itself. David Talmage[5] and Macfarlane Burnet[6] recognized that this theory would make more sense if the postulated natural antibodies were located on the surface of what we now call B lymphocyte cells. If each cell were endowed with only one sort of antibody specificity, then the antigen could select one lymphocyte out of a repertoire, cause its clonal division and stimulate antibody production and secretion (Fig. 1). In 1958, Joshua Lederberg[7] and I provided the first evidence for the clonal selection theory, namely that one B cell always produces only one antibody.

DNA shuffling in antibody formation
Antibodies are multichain proteins that come in different forms. The most abundant, immunoglobulin-γ (IgG), consists of two identical light (L) chains and two identical heavy (H) chains[8] (Fig. 2a). The carboxy-terminal halves of the two light chains are identical to each other, but the amino-terminal halves differ in more than 50 residues, called the 'variable' region[9]. The heavy chains, too, consist of a variable (V) part and a constant (C) part.

In 1965, William J. Dreyer and J. Claude Bennett[10] bucked the dogma at the time that 'one gene makes one protein', and put forward the revolutionary concept that the carboxy-terminal C region of the L chain was always encoded by a single gene, but that the amino-terminal V half could be encoded by multiple separate genes, perhaps as many as 100,000 in number. It followed that a chosen V gene must then somehow become associated with a C gene by a DNA rearrangement event in each lymphocyte cell, because only when the V–C regions were spliced together could a functional protein be expressed.

At that time, there was no way to interrogate the genome directly to test this concept, and for a decade debate and controversy raged. One school, led by Leroy Hood, favoured the idea that a large array of germline-encoded V genes for the L and H chains underwent rearrangement. At the other extreme were proponents of a single, very highly mutable V gene that was extensively mutated in emerging B cells. In the middle were those who favoured the idea of a handful of V genes that were subject to extensive recombination in somatic cells (the cells of the body, excluding the sex cells). This compromise was supported by luminaries such as Oliver Smithies and Gerald Edelman. Francis Crick was quite taken with the idea that just two V genes undergo rearrangement in the germ line, with further mutation in somatic cells.

Before arriving at the solution brought by advances in molecular biology, one more fact is worthy of note. Elvin Kabat astutely pointed out that in the V regions of both heavy and light chains there were also three short stretches of amino acids where variation was considerably greater than elsewhere in the molecules, and these so-called hyper-variable regions were deemed likely to be sites of union with the antigen[11]. Could it be that there was actually an assembly of several, rather than just two, genes encoding each chain?

The new tools for manipulating and sequencing DNA came to the rescue. In 1976, Nobumichi Hozumi and Susumu Tonegawa[12] conducted a landmark experiment. They used a DNA cutting enzyme known as a restriction endonuclease to digest the DNA extracted from a mouse embryo and from an antibody-secreting tumour. The resulting DNA fragments were then separated on the basis of size, and

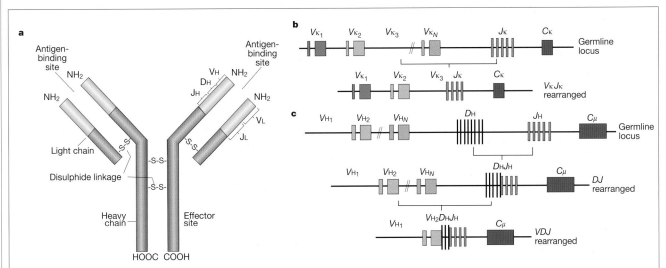

Figure 2 Antibody formation. **a**, Structure of an antibody (immunoglobulin). Two identical heavy chains are connected by disulphide linkages. The antigen-recognizing site is composed of the variable regions (yellow) of the heavy and light chains, whereas the effector site (which determines its function) is determined by the amino-acid sequence of the heavy chain constant region (red). **b**, Assembly of the light (κ)- and heavy (H)-chain genes of antibodies by somatic recombination during B-lymphocyte development. The L chain is encoded by variable (V), joining (J) and constant (C) genes. While the developing B cell is still maturing in the bone marrow, one of the 30–40 V genes combine with one of the five J genes and is juxtaposed to a C gene. The recombining process involves deletion of the intervening DNA between the selected genes. **c**, The H chain is encoded by V, D, J and C genes. The assembly of the H chain gene occurs in two stages: one of the D genes joins with a J gene, then one of the V genes joins with that DJ assembly.

were reacted with radioactive probes, one corresponding to the whole L chain, the other to just the C portion. In the tumour, both probes lit up the identical fragment, whereas in the embryonic extract, two different fragments hybridized to the full-length probe, but only one of them to the C-region probe. Both fragments from the embryo sample were different in size from the single hybridizing fragment in the tumour sample. The experiment strongly argued for the V and C genes being some distance from each other in the embryo, but having been rearranged and assembled during development of antibody-forming cells in adults to form a continuous DNA sequence constituting the full L-chain gene.

Generating more diversity

Definitive elucidation of immunoglobulin gene structure depended on molecular cloning and subsequent sequencing of the genes themselves[13]. Here came another surprise, and one that could really not have been anticipated. V-region genes in the germline were found to be significantly shorter than is required to code for the V region of the L chain. It turned out that there is a series of 'minigenes' known as 'joining' (J) genes, which code for about 13 amino acids of the L chain. Thus, the full L chain is actually encoded by V, J and C genes (Fig. 2b). For the H chain, it is still more complicated, as there exists a series of 'diversity' (D) genes that encode up to eight amino acids that lie between the V and J regions. Thus, the H chain is encoded by V, D, J and C genes (Fig. 2c). The assembly of a complete H-chain V region occurs in two separate steps: first, one of the D regions joins with one of the J regions, then one of many V regions joins with that DJ assembly (Fig.2c). The joining process is followed by deletion of the intervening DNA between the chosen minigenes. This is the first example of a somatic cell possessing a different genome from its fellow cells.

This minigene assembly process has important implications for antibody diversity. In humans, here are two types of L chains, κ and λ, each with its own sets of V and J genes. For the κ light chain, there are 40 functional V genes and 5 functional J genes; for the λ chain, there are 31 and 4, respectively. There is only one kind of variable region for the H chain, encoded by 51 V genes, 25 D genes and 6 J genes. To a first approximation, therefore, there are $(40 \times 5) + (31 \times 4) = 324$ different possible assemblies of L chains, and $51 \times 25 \times 6 = 7,650$ combinations for H chains. Thus, together, there are potentially 2,478,600 different types of germline-encoded antibodies.

But this is a considerable underestimate for two reasons. Recombination junctions can occur at different positions and this junctional diversity increases variability. Furthermore, a few extra nucleotides, called N regions, can be inserted between D–J junctions and V–D junctions in many H chains, and in a smaller percentage of L-chain V–J junctions. These nucleotides are not present in the germline and add to antibody diversity.

Yet further diversity can be generated by DNA mutations in dividing B cells. B cells expressing newly assembled immunoglobulin genes, each with its own unique specificity, constitute the 'primary repertoire'. When an antigen stimulates a chosen B cell to divide (Fig. 1), a proportion of the progeny migrate into the vicinity of antigen-capturing follicular dendritic cells (FDCs) and gradually form a 'germinal centre'. FDCs retain antigen on their surface for long periods and stimulate further rounds of division. Within the germinal centre the B cells display an extraordinarily high rate of somatic mutation in V genes, estimated at 10^{-3} per nucleotide per division[14]. As antibody production accumulates, only those B cells with heightened affinity for the antigen gain access to FDC-bound antigen and thus are further stimulated to divide. As a result, B-cell clones secreting higher affinity antibody are selected in an iterative manner.

The 'memory' B cells that emerge from the germinal centre constitute the 'secondary repertoire', which is even more diverse than the primary one. Twenty mutations per chain are not uncommon; nor are thousand-fold increases in affinity. Thus, as it turned out, two early theories of antibody diversification proved to be correct: rearrangement of germline genes gives the naive B-cell

repertoire, and somatic mutation ensures further diversification during memory B-cell development.

Switching function

There are several different classes of antibodies, all of which have distinct roles that are also produced by rearrangements at the DNA level. There are eight different genes for the C region of the H chain, which specify different antibody functions. Each B cell first links the chosen VDJ assembly to a C gene known as μ, creating an antibody class called IgM. If that cell is propelled into a pathway favouring the production of an antibody prominent in mucus secretions (such as in the gastrointestinal tract), the VDJ section is switched over to a C region encoded by C gene α, and the cell produces IgA. If, on the other hand, the antigen is of parasite origin, or an allergen such as a pollen grain, the cell may be stimulated to produce IgE, in which case the VDJ region associates with the product of the C gene ε. All of this occurs without any change in the specificity of the antibody being secreted. Although the detailed molecular mechanisms are still being investigated, the class switching again involves sequential excision of portions of the genetic material. Cytidine deaminase induced by B-cell-specific activation may be significant in both class switching and somatic hypermutation.

Assembly of T-cell receptors

Whereas B cells make antibodies against antigens, the thymus-derived or T cells also respond to foreign agents, specializing in a more localized form of combat. Cytotoxic T cells are capable of killing virus-infected cells or cells displaying cancer-specific antigens. Other

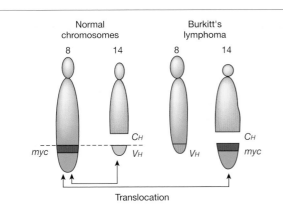

Figure 3 Reciprocal chromosomal translocations in Burkitt's lymphoma, a solid tumour of B lymphocytes. The genes for making the heavy chains of antibodies (C$_H$) are located on chromosomes 14, whereas those for making the light chains are on chromosomes 2 and 22. These genes are expressed exclusively in B lymphocytes, because only these cells have the necessary transcription factors to switch on their expression. In most (over 90%) of Burkitt's lymphoma cases, a reciprocal translocation moves the proto-oncogene c-*myc* from its normal position on chromosome 8 to a location close to the antibody heavy-chain genes on chromosome 14 (ref. 18). In other cases, c-*myc* is translocated close to the antibody genes on chromosome 2 or 22. In every case, c-*myc* now finds itself in a region of active gene transcription, and it may simply be the overproduction of the c-*myc* product (a transcription factor essential for cell division) that propels the lymphocyte down the pathway towards cancer.

T cells secrete powerful stimulatory and inflammatory molecules, most of which act in a strictly localized context. T cells can also help guide B cells down appropriate pathways of differentiation.

In common with B cells, T cells also have one-receptor specificity and, for simplicity, I shall mention only the αβ T-cell receptor (TCR), a heterodimer consisting of two subunits, the α- and β-chains, joined by disulphide bonds[15,16]. The strategy for generating T cells with different receptors is strikingly similar to that used by B cells to produce different antibodies and, in fact, the TCR binding surface looks much like that of an antibody. The β-chain of the TCR is assembled in somatic cells from V, D, J and C genes; the α-chain from V, J and C genes. There are additions of N-region nucleotides between V and D, as well as between D and J on the β-chain; and between V and J on the α-chain. A similar rearrangement also takes place for the γδ TCR.

But something peculiar about T-cell recognition was noted by Rolf Zinkernagel and Peter Doherty[17], who demonstrated that cytotoxic T cells could recognize viral antigens only if a specific 'self' molecule were also present on the target cell (see Box 1). The key part of the T-cell recognition puzzle fell into place when it was discovered that the TCR recognized short antigenic peptides bound to the groove of a self molecule known as the major histocompatibility complex (MHC), as well as surrounding portions of the MHC molecule itself. Cells have special mechanisms for fragmenting proteins into peptides of 8–24 amino acids in length, attaching these to MHC molecules and transporting the entire complex to the cell surface. TCRs then 'see' these short linear portions of antigens, be these of viral, bacterial or parasitic origin, or even portions of normal intracellular components. Such a system can help to control infections where the pathogen goes 'underground' inside a cell, and can also eliminate cells with mutated self antigens, such as cancer cells.

Lymphocytes and cancer

Lymphocytes have been a favourite tool in cancer research. A notable example of DNA science applied in this way relates to the B-cell tumour of humans known as Burkitt's lymphoma. Occasionally DNA strands break and are incorrectly repaired. Thus, a piece of a chromosome becomes attached to the broken end of another one, and vice versa, in a process known as reciprocal translocation (Fig. 3).

In the case of Burkitt's lymphoma, a tumour-promoting gene or oncogene called *myc* is translocated from its normal position on chromosome 8 right into the middle of the IgH chain locus on chromosome 14 (ref. 18). In this highly active transcriptional environment, *myc* expression is switched on, and eventually cancer develops.

It has been possible to create lymphoma-prone transgenic mice, which express *myc* in aberrantly high amounts. Because cancer is typically a multistage process, if further oncogenes are expressed simultaneously in transgenic mice, the onset of cancer can be dramatically accelerated. One such example involves the gene *bcl-2*. When this gene is expressed, it stops cells from undergoing natural programmed death (apoptosis)[19]. Mice expressing *myc* and *bcl-2* showed very rapid development of tumours. An enormous amount of literature has accumulated related to the expanding family of *bcl-2*-related genes and their roles in the regulation of programmed cell death. Models derived from lymphocytes and their malignancies have led to insights with implications well beyond immunology.

DNA vaccines

DNA research has been of immense value to vaccine research. Through gene cloning and expression, candidate antigens can be identified and tested. In an era of rapid nucleotide sequencing, the whole genome of a pathogen can be determined, and computer programs can search for sequences likely to encode outer membrane proteins, which can be assessed as candidate vaccine molecules (see, for example, the series of papers published recently in *Nature* (**419**, 489–542, 2002) on the genomics of the malaria parasite).

Amazingly, DNA itself can serve as a vaccine. DNA vaccines work on the principle that the gene sequence for one or more candidate antigens is introduced into an animal or person via a delivery vehicle known as a vector, together with a strong promoter that can switch on its expression in mammalian cells. Cells that take up the injected DNA transcribe and translate the gene and release the relevant antigen protein, which the body can in turn manufacture antibodies against. Thus, the body itself becomes a vaccine factory[20]. Unfortunately, so far this approach has worked better in mice than in humans, but many avenues are being pursued to improve this situation.

To strengthen the immune response to a vaccine, it may be necessary to use an adjuvant substance. Here, DNA may also be of potential use. Scavenger cells, which capture antigens, have evolutionarily conserved receptors, known as Toll-like receptors (TLRs), which recognize antigens common to many pathogens. One such receptor is TLR-9, which recognizes unmethylated CpG motifs commonly found in bacterial but not mammalian DNA. Accordingly, unmethylated CpG-rich DNA sequences represent a promising new category of adjuvant[21].

Future directions

The solutions to the puzzle of antibody diversity and mystery of T-cell recognition of antigenic peptides are among the brightest chapters of biology in the last quarter of the twentieth century. The future of immunology will be all about how the system is regulated and how it makes decisions: whether to respond or not; whether to direct efforts towards antibody formation or cell-mediated immunity; and, if the latter, whether more towards cytokine-secreting T cells or cytotoxic T cells.

As in the past, the future will be about information and thus about DNA science. All the complex signalling pathways, the feedback loops, and the intricate rules governing cell division on the one hand or programmed cell death on the other, will be progressively revealed. As this happens, the possibilities for applied research and development will be immense. In particular, new therapeutic targets will be identified. The 'miracle' drug for chronic myelogenous leukaemia, Glivec, was made possible after the characterization of the extraordinary cancerous potential of the chimaeric oncogene *bcr–abl*. This will surely be only the first of a plethora of more intelligently designed anti-cancer drugs. Potent cytokines and monoclonal antibodies directed against cell surface-associated structures are already prominent within a radically revised pharmaceutical armamentarium in areas including cancer, autoimmunity, allergy and transplantation. DNA research is therefore crucial to a new generation of immunologists, from those striving towards the development of novel vaccines to those seeking to understand and control autoimmune diseases, allergy and transplant tolerance.

doi:10.1038/nature01409

1. Ehrlich, P. The Croonian Lecture. On immunity with special reference to cell life. *Proc. R. Soc. Lond. B* **66**, 424–448 (1900).
2. Landsteiner, K. *The Specificity of Serological Reactions* (Harvard Univ. Press, Boston, 1945).
3. Breinl, F. & Haurowitz, F. Untersuchungen des Präzipitates aus Hämoglobin und anti-Hämoglobin-Serum und Bemerkungen über die Natur der Antikörper. *Hoppe-Seyler's Z. Physiol. Chem.* **192**, 45–57 (1930).
4. Jerne, N. K. The natural selection theory of antibody formation. *Proc. Natl Acad. Sci. USA* **41**, 849–857 (1955).
5. Talmage, D. W. Allergy and immunology. *Annu. Rev. Med.* **8**, 239–256 (1957).
6. Burnet, F. M. A modification of Jerne's theory of antibody production using the concept of clonal selection. *Aust. J. Sci.* **20**, 67–69 (1957).
7. Nossal, G. J. V. & Lederberg, J. Antibody production by single cells. *Nature* **181**, 1419–1420 (1958).
8. Edelman, G. M. & Gall, W. E. The antibody problem. *Annu. Rev. Biochem.* **38**, 415–466 (1969).
9. Hilschmann, N. & Craig, L. C. Amino acid sequence studies with Bence-Jones proteins. *Proc. Natl Acad. Sci. USA* **53**, 1403–1409 (1965).
10. Dreyer, W. J. & Bennett, J. C. The molecular basis of antibody formation: a paradox. *Proc. Natl Acad. Sci. USA* **54**, 864–869 (1965).
11. Wu, T. T. & Kabat, E. A. An analysis of the sequences of the variable regions of Bence Jones proteins and myeloma light chains and their implications for antibody complementarity. *J. Exp. Med.* **132**, 211–250 (1970).
12. Hozumi, N. & Tonegawa, S. Evidence for somatic rearrangement of immunoglobulin genes coding for variable and constant regions. *Proc. Natl Acad. Sci. USA* **73**, 3628–3632 (1976).
13. Bernard, O., Hozumi, N. & Tonegawa, S. Sequences of mouse immunoglobulin light chain genes before and after somatic changes. *Cell* **15**, 1133–1144 (1978).
14. Kocks, C. & Rajewsky, K. Stable expression and somatic hypermutation of antibody V regions in B-cell developmental pathways. *Annu. Rev. Immunol.* **7**, 537–559 (1989).
15. Hedrick, S. M., Cohen, D. I., Nielsen, E. A. & Davis, M. M. Isolation of cDNA clones encoding T cell-specific membrane-associated proteins. *Nature* **308**, 149–153 (1984).
16. Yanagi, Y. *et al.* A human T cell-specific cDNA clone encodes a protein having extensive homology to immunoglobulin chains. *Nature* **308**, 145–149 (1984).
17. Zinkernagel, R. M. & Doherty, P. C. Restriction of *in vitro* T cell-mediated cytotoxicity in lymphocytic choriomeningitis within a syngeneic or semiallogeneic system. *Nature* **248**, 701–702 (1974).
18. Bernard, O., Cory, S., Gerondakis, S., Webb, E. & Adams, J. M. Sequence of the murine and human cellular myc oncogenes and two modes of myc transcription resulting from chromosome translocation in B lymphoid tumours. *EMBO J.* **2**, 2375–2383 (1983).
19. Vaux, D. L., Cory, S. & Adams, J. M. Bcl-2 gene promotes haemopoietic cell survival and cooperates with c-*myc* to immortalize pre-B cells. *Nature* **335**, 440–442 (1988).
20. Donnelly, J. J., Ulmer, J. B., Shiver, J. W. & Liu, M. A. DNA vaccines. *Annu. Rev. Immunol.* **15**, 617–648 (1997).
21. Krieg, A. M. CpG motifs in bacterial DNA and their immune effects. *Annu. Rev. Immunol.* **20**, 709–760 (2002).
22. Burnet, F. M. & Fenner, F. J. *The Production of Antibodies* (Macmillan, Melbourne, 1949).
23. Billingham, R. E., Brent, L. & Medawar, P. B. Actively acquired tolerance of foreign cells. *Nature* **172**, 603–606 (1953).
24. Nossal, G. J. V. & Pike, B. L. Clonal anergy: persistence in tolerant mice of antigen-binding B lymphocytes incapable of responding to antigen or mitogen. *Proc. Natl Acad. Sci. USA* **77**, 1602–1606 (1980).

Original reference: *Nature* **421**, 440–444 (2003).

The digital code of DNA

Leroy Hood* **& David Galas†**

*Institute for Systems Biology, 4225 Roosevelt Way NE, Seattle, Washington 98105, USA (e-mail: lhood@systemsbiology.org)
†Keck Graduate Institute of Applied Sciences, 535 Watson Drive, Claremont, California 91711, USA (e-mail: david_galas@kgi.edu)

The discovery of the structure of DNA transformed biology profoundly, catalysing the sequencing of the human genome and engendering a new view of biology as an information science. Two features of DNA structure account for much of its remarkable impact on science: its digital nature and its complementarity, whereby one strand of the helix binds perfectly with its partner. DNA has two types of digital information — the genes that encode proteins, which are the molecular machines of life, and the gene regulatory networks that specify the behaviour of the genes.

"Any living cell carries with it the experiences of a billion years of experimentation by its ancestors." Max Delbruck, 1949.

The discovery of the double helix in 1953 immediately raised questions about how biological information is encoded in DNA[1]. A remarkable feature of the structure is that DNA can accommodate almost any sequence of base pairs — any combination of the bases adenine (A), cytosine (C), guanine (G) and thymine (T) — and, hence any digital message or information. During the following decade it was discovered that each gene encodes a complementary RNA transcript, called messenger RNA (mRNA)[2], made up of A, C, G and uracil (U), instead of T. The four bases of the DNA and RNA alphabets are related to the 20 amino acids of the protein alphabet by a triplet code — each three letters (or 'codons') in a gene encodes one amino acid[3]. For example, AGT encodes the amino acid serine. The dictionary of DNA letters that make up the amino acids is called the genetic code[4]. There are 64 different triplets or codons, 61 of which encode an amino acid (different triplets can encode the same amino acid), and three of which are used for 'punctuation' in that they signal the termination of the growing protein chain.

The molecular complementary of the double helix — whereby each base on one strand of DNA pairs with its complementary base on the partner strand (A with T, and C with G) — has profound implications for biology. As implied by James Watson and Francis Crick in their landmark paper[1], base pairing suggests a template-copying mechanism that accounts for the fidelity in copying of genetic material during DNA replication (see accompanying article by Alberts, page 117). It also underpins the synthesis of mRNA from the DNA template, as well as processes of repairing damaged DNA (discussed by Friedberg, page 122).

Tools to modify DNA

The enzymes that function in cells to copy, cut and join DNA molecules were also exploited as key tools for revolutionary new techniques in molecular biology, including the cloning of genes and expression of their proteins, and mapping the location of genes on chromosomes. The ability to recreate the process of DNA replication artificially in the laboratory led to the development of two techniques

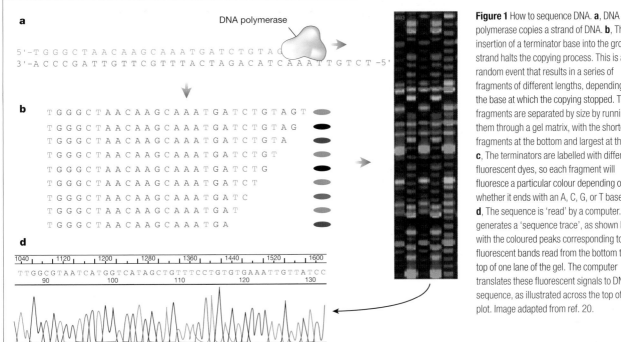

Figure 1 How to sequence DNA. **a**, DNA polymerase copies a strand of DNA. **b**, The insertion of a terminator base into the growing strand halts the copying process. This is a random event that results in a series of fragments of different lengths, depending on the base at which the copying stopped. The fragments are separated by size by running them through a gel matrix, with the shortest fragments at the bottom and largest at the top. **c**, The terminators are labelled with different fluorescent dyes, so each fragment will fluoresce a particular colour depending on whether it ends with an A, C, G, or T base. **d**, The sequence is 'read' by a computer. It generates a 'sequence trace', as shown here, with the coloured peaks corresponding to fluorescent bands read from the bottom to the top of one lane of the gel. The computer translates these fluorescent signals to DNA sequence, as illustrated across the top of the plot. Image adapted from ref. 20.

that transformed biology: a manual DNA sequencing method in 1975 and, in 1985, the discovery of the polymerase chain reaction (PCR), whereby DNA sequences could be amplified a millionfold or more[5].

Although sequencing and PCR transformed the science of biology, they also had wide applications for medicine and forensics. The detection of variation in DNA sequence from one individual to the next — so-called 'polymorphisms' — forms the basis of DNA 'finger-printing' of individuals. Forensics uses these fingerprints to deal with paternity disputes, as well as criminal cases such as rape. The finding that many specific DNA polymorphisms are associated with disease or disease susceptibility has brought DNA diagnostics to medicine and opened the pathway to truly predictive medicine, where the risks of disease can be identified in advance of symptoms (see accompanying article by Bell, page 100).

Automated DNA sequencing

The first efforts to sequence DNA, pioneered by Walter Gilbert[6] and Fred Sanger[7] in the 1970s, decoded stretches of DNA a few hundred bases long. When the first complete genome was sequenced over a period of about one year in 1977–78 — that of a viral genome of about 5,000 bases[8] — it became clear that DNA sequence data could provide unique insights into the structure and function of genes, as well as genome organization. It was this potential to generate vast amounts of information about an organism from its genetic code that inspired efforts towards the automation of DNA sequencing (Fig. 1).

The combination of technical wizardry and intensive automation in the decade that followed launched the 'genomic era'. A series of new instruments enabled novel approaches to biological analysis[9–11]. The first sequencing machine — invented by Leroy Hood, Lloyd Smith and Mike Hunkapiller in 1986 (ref. 12) — was automated in data acquisition, but still required substantial manual attention and the sequencing rate was low, roughly 250 bases per day. Over the next ten years, the development of automated DNA sequencing accelerated, rapidly passing through three distinct stages: the prototype sequencing machine (1986); a robust instrument that could be used routinely in a standard laboratory (1989); and finally, a machine that formed part of an integrated factory-like production line where DNA sample preparation and sequencing were all fully automated (1998). The

advances in sequencing capacity have been striking — the latest sequencing machines are able to decode approximately 1.5 million bases over 24 hours — 6,000 times the throughput of the prototype.

The goals of high-throughput biological instrumentation are to increase throughput, enhance the quality of the data, and greatly reduce the cost of per unit information acquired. To reach these goals in the future, the miniaturization, automation, parallelization and integration of successive procedures will propel DNA sequencing technology into the realm of microfluidics and microelectronics, and eventually into the area of nanotechnology. With single-DNA-molecule sequencing, we foresee a time when the entire genome of an individual could be sequenced in a single day at a cost of less than $US10,000 (compared with the US$50 million or more it would cost today). This will readily enable the decoding of the genomic sequence of virtually any organism on the planet and provide unparalleled access to the foundations of biology and the study of human genetic variability.

The Human Genome Project

The breathtaking speed at which automated DNA sequencing developed was largely stimulated by the throughput demands of the Human Genome Project (HGP), which officially started in 1990 following discussions and studies on feasibility and technology that began in earnest in 1985. The objectives of the HGP were to generate a finished sequence in 15 years[13], but a draft of the human genome sequence was available in 2001. Two versions of the draft were generated and published in 2001, one by the publicly funded International Human Genome Sequencing Consortium[14], and another by the biotechnology company Celera[15] (Box 1). In the process of developing the tools and methodology to be able to sequence and assemble the 3 billion bases of the human genome, a range of plant, animal and microbial genomes was sequenced and many more are currently being decoded. As genome sequences become available, different areas of biology are being transformed — for example, the discipline of microbiology has changed significantly with the completion of more than 100 bacterial genome sequences over the past decade.

The HGP profoundly influenced biology in two respects. First, it illustrated the concept of 'discovery science' — the idea that all the elements of the system (that is, the complete genome sequence and the entire RNA and protein output encoded by the genome) can be defined, archived in a database, and made available to facilitate hypothesis-driven science and global analyses. Second, to succeed, the HGP pushed the development of efficient large-scale DNA sequencing and, simultaneously, drove the creation of high-through-put tools (for example, DNA arrays and mass spectrometry) for the analysis of other types of related biological information, such as mRNAs, proteins and molecular interactions.

The digital nature of biological information
The value of having an entire genome sequence is that one can initiate the study of a biological system with a precisely definable digital core of information for that organism — a fully delineated genetic source code. The challenge, then, is in deciphering what information is encoded within the digital code. The genome encodes two main types of digital information — the genes that encode the protein and RNA molecular machines of life, and the regulatory networks that specify how these genes are expressed in time, space and amplitude.

It is the evolution of the regulatory networks and not the genes themselves that play the critical role in making organisms different from one another. The digital information in genomes operates across three diverse time spans: evolution (tens to millions of years), development (hours to tens of years), and physiology (milliseconds to weeks). Development is the elaboration of an organism from a single cell (the fertilized egg) to an adult (for humans this is 10^{14} cells of thousands of different types). Physiology is the triggering of specific functional programmes (for example, the immune response) by environmental cues. Regulatory networks are crucial in each of these aspects of biology.

Regulatory networks are composed of two main types of components: transcription factors and the DNA sites to which they bind in the control regions of genes, such as promoters, enhancers and silencers. The control regions of individual genes serve as information processors to integrate the information inherent in the concentrations of different transcription factors into signals that mediate gene expression. The collection of the transcription factors and their cognate DNA-binding sites in the control regions of genes that carry out a particular developmental or physiological function constitute these regulatory networks (Fig. 2).

Because most 'higher' organisms or eukaryotes (organisms that contain their DNA in a cellular compartment called the nucleus), such as yeast, flies and humans, have predominantly the same families of genes, it is the reorganization of DNA-binding sites in the control regions of genes that mediate the changes in the developmental programmes that distinguish one species from another. Thus, the regulatory networks are uniquely specified by their DNA-binding sites and, accordingly, are basically digital in nature.

One thing that is striking about digital regulatory networks is that they can change significantly in short periods of evolutionary time. This is reflected, for example, in the huge diversity of the body plans, controlled by gene regulatory networks, that emerged over perhaps 10–30 million years during the Cambrian explosion of metazoan organisms (about 550 million years ago). Likewise, remarkable changes occurred to the regulatory networks driving the development of the human brain during its divergence from its common ancestor with chimpanzees about 6 million years ago.

Biology has evolved several different types of informational hierarchies. First, a regulatory hierarchy is a gene network that defines the relationships of a set of transcription factors, their DNA-binding sites and the downstream peripheral genes that collectively control a particular aspect of development. A model of development in the sea urchin represents a striking example[16] (Fig. 2). Second, an evolutionary hierarchy defines an order set of relationships, arising from DNA duplication. For example, a single gene may be duplicated to generate a multi-gene family, and a multi-gene family may be duplicated to create a supergene family. Third, molecular machines may be assembled into structural hierarchies by an ordered assembly process. One

Box 1
Sequencing the human genome

The first complete drafts of the human genome sequence were published in 2001 by the International Human Genome Sequencing Consortium (IHGSC), a publicly funded effort, and Celera, a biotechnology company, using different approaches. Both efforts used a random or shotgun approach where the original DNA to be sequenced was randomly broken into overlapping fragments that were then cloned, and 500 base pairs (bp) were 'read' from one or both ends of the clones.

For the draft genome sequences, each base was read six to ten times to optimize the accuracy of the sequence. The stretches of DNA sequence were read by a computer and assembled into a complete sequence. The IHGSC effort randomly cleaved DNA into ~ 200,000-bp fragments and generated a map of these fragments across the 24 different human chromosomes; it then used the shotgun approach to sequence the pre-ordered fragments clone by clone. In contrast, Celera randomly fragmented the entire genome into three sizes of fragments (approximately 2,000, 10,000 and 200,000 bp), sequenced both ends of the clones and then used the end sequences to assemble the entire genome sequence, without the aid of a map.

Celera's 1998 announcement that it would sequence the human genome within three years was greeted with considerable scepticism, but it succeeded in producing a draft sequence and considerably accelerating the public effort. The efforts of both groups benefited science by producing draft genome sequences considerably earlier than expected.

Although minor differences were noted between the two drafts, the overall conclusions concerning gene numbers, repeated sequences and chromosomal organization were remarkably similar. For example, both groups identified 30,000–35,000 genes, far fewer than the 100,000 expected from an earlier (admittedly 'back of the envelope') calculation.

example of this is the basic transcription apparatus that involves the step-by-step recruitment of factors and enzymes that will ultimately drive the specific expression of a given gene. A second example is provided by the ribosome, the complex that translates RNA into protein, which is assembled from more than 50 different proteins and a few RNA molecules. Finally, an informational hierarchy depicts the flow of information from a gene to environment: gene → RNA → protein → protein interactions → protein complexes → networks of protein complexes in a cell → tissues or organs → individual organisms → populations → ecosystems. At each successively higher level in the informational hierarchy, information can be added or altered for any given element (for example, by alternative RNA splicing or protein modification).

Systems approaches to biology
Humans start life as a single cell — the fertilized egg — and develop into an adult with trillions of cells and thousands of cell types. This process uses two types of biological information: the digital information of the genome, and environmental information, such as metabolite concentrations, secreted or cell-surface signals from other cells or chemical gradients. Environmental information is of two distinct types: deterministic information where the consequences of the signals are essentially predetermined, and stochastic information where chance dictates the outcome.

Random, or stochastic, signals can generate significant noise in biological systems, but it is only in special cases that noise is converted into signals. For example, stochastic events govern many of the genetic mechanisms responsible for generating antibody diversity. In the immune response, those B cells that produce antibodies that bind

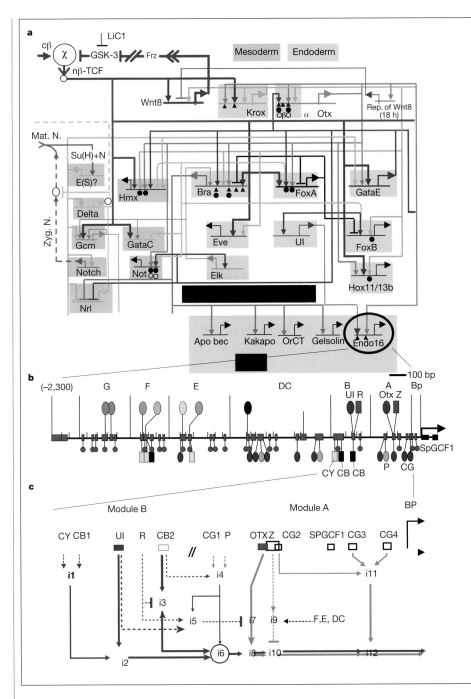

Figure 2 A gene regulatory network involved in sea urchin development[16]. **a**, Part of the network of transcription factors and their interactions with the control regions of other transcription factors. Genes are indicated by horizontal lines; arrowheads indicate activation; '⊥' symbols indicate gene repression. **b**, An enlargement of the promoter region of a gene, called *endo 16*, that helps modulate the development of the endoderm. It contains 34 binding sites (rectangles) for 13 different transcription factors and cofactors (illustrated as rectangles or lollipops, respectively). Six modules (A–G) of transcription factors and binding sites carry out discrete functions to developmentally regulate *endo 16*. **c**, Diagram depicting the logical structures of the A and B control circuits during sea urchin development.

tightly to the antigen (that is, those having high affinities) undergo an expansion in number that is proportional to the strength of the antibody affinity (see accompanying article by Nossal, page 126). Hence, the signal (high affinity) is distinguished from the noise (low affinity). Moreover, high levels of mutation in the B cells causes specific diversification of antibody genes in the presence of antigen and permits the affinity to increase even more. The cells carrying the higher-affinity antibody genes are then preferentially selected for survival and proliferation.

The key question is what and how much signal emerges from the noise. Analysis of stochastic events and the differentiation between signal and noise will be a future challenge for contemporary biology. The immune response has been studied for more than 100 years, yet we still have only a partial understanding of its systems properties, such as the immune response and tolerance (the unresponsiveness to one's own cells). This is because until recently immunologists have been able to study this complex system only one gene or one protein at a time.

The systems approach permits the study of all elements in a system in response to genetic (digital) or environmental perturbations. Global quantitative analyses of biological information from different levels each provide new insights into the operation of the system; hence, information at as many levels as possible must be captured, integrated, and ultimately, modelled mathematically. The model should explain the properties of the system and establish a framework that allows us to redesign the system in a rational way to generate new emergent properties.

Several systems have been explored successfully. The utilization of the sugar galactose in yeast has been analysed using genetic perturbations (inactivation of genes) and four levels of information were gathered — RNA and protein concentrations as well as protein–protein and protein–DNA interactions[17]. Using an iterative and integrative systems approach, new insights into the regulation of galactose use were gained. Moreover, the relationships of the galactose regulatory network to other modules in the yeast cell were also delineated. Likewise, systems approaches to early embryonic development in the sea urchin have delineated a regulatory network that has significant predictive power[16] (Fig. 2). Finally, systems approaches to metabolism in an archaeal halobacterium (an organism thriving in up to five-molar salt solutions, such as the Dead Sea) have revealed new insights into the inter-relationships among several modules controlling energy production in the cell[18].

The study of cellular and organismal biology using the systems approach is at its very beginning. It will require integrated teams of scientists from across disciplines — biologists, chemists, computer scientists, engineers, mathematicians and physicists. New methods for acquiring and analysing high-throughput biological data are needed. A powerful computational infrastructure must be leveraged to generate more effective approaches to the capture, storage, analysis, integration, graphical display and mathematic formulation of biological complexity. New technologies must be integrated with each other. Finally, hypothesis-driven and discovery science must be integrated. In short, both new science and technology must emerge for the systems biology approach to realize its promise. A cultural shift in the biological sciences is needed, and the education and training of the next generation of biologists will require significant reform.

Gordon Moore, the founder of Intel, predicted that the number of transistors that could be placed on a computer chip would double every 18 months. It has for more than 30 years. This exponential growth has been a driver for the explosive growth of information technology. Likewise, the amount of DNA sequence information available to the scientific community is following a similar, perhaps even steeper, exponential increase. The critical issue is how sequence information can be converted into knowledge of the organism and how biology will change as a result. We believe that a systems approach to biology is the key. It is clear, however, that this approach poses significant challenges, both scientific and cultural[19]. The discovery of DNA structure started us on this journey, the end of which will be the grand unification of the biological sciences in the emerging, information-based view of biology. □

doi:10.1038/nature01410

1. Watson, J. D. & Crick, F. H. C. A structure for deoxyribose nucleic acid. *Nature* **171**, 737–738 (1953).
2. Brenner, S., Jacob, F. & Meselson, M. An unstable intermediate carrying information from genes to ribosomes for protein synthesis. *Nature* **190**, 576–581 (1961).
3. Crick, F. H. C., Barnett, L., Brenner, S. & Watts-Tobin, R. J. General nature of the genetic code for proteins. *Nature* **192**, 1227–1232 (1961).
4. Nirenberg, M. W. & Matthaei, J. H. The dependence of cell-free protein synthesis in *E. coli* upon naturally occurring or synthetic polynucleotides. *Proc. Natl Acad. Sci. USA* **47**, 1588–1602 (1961).
5. Saiki, R. K. *et al.* Enzymatic amplification of β-globin sequences and restriction site analysis for diagnosis of sickle cell anemia. *Science* **230**, 1350–1354 (1985).
6. Maxam, A. M. & Gilbert, W. A new method of sequencing DNA. *Proc. Natl Acad. Sci. USA* **74**, 560–564 (1977).
7. Sanger, F. & Coulson, A. R. A rapid method for determining sequences in DNA by primed synthesis with DNA polymerase. *J. Mol. Biol.* **94**, 444–448 (1975).
8. Sanger, F. *et al.* Nucleotide sequence of bacteriophage φX174. *Nature* **265**, 678–695 (1977).
9. Hunkapiller, M. W. & Hood, L. New protein sequenator with increased sensitivity. *Science* **207**, 523–525 (1980).
10. Horvath, S. J., Firca, J. R., Hunkapiller, T., Hunkapiller M. W. & Hood. L. An automated DNA synthesizer employing deoxynucleoside 3′ phosphoramidites. *Methods Enzymol.* **154**, 314–326 (1987).
11. Kent, S. B. H., Hood, L. E., Beilan, H., Meister S. & Geiser, T. High yield chemical synthesis of biologically active peptides on an automated peptide synthesizer of novel design. *Peptides* **5**, 185–188 (1984).
12. Smith, L. M. *et al.* Fluorescence detection in automated DNA sequence analysis. *Nature* **321**, 674–679 (1986).
13. Collins, F. & Galas, D. J. A new five-year plan for the US Human Genome Project. *Science* **262**, 43–46 (1993).
14. Venter, J. C. *et al.* The sequence of the human genome. *Science* **291**, 1304–1351 (2001).
15. International Human Genome Sequencing Consortium. Initial sequencing and analysis of the human genome. *Nature* **409**, 860–921 (2001).
16. Davidson, E. H. *et al.* A genomic regulatory network for development. *Science* **295**, 1669–1678 (2002).
17. Ideker, T. *et al.* Integrated genomic and proteomic analyses of a systematically perturbed metabolic network. *Science* **292**, 929–933 (2001).
18. Baliga, N. S. *et al.* Coordinate regulators of energy transduction modules in the *Halobacterium* sp. analyzed by a global systems approach. *Proc. Natl Acad. Sci. USA* **99**, 14913–14918 (2002).
19. Aderem A. & Hood, L. Immunology in the post-genomic era. *Nature Immunol.* **2**, 1–3 (2001).
20. Dennis, C. & Gallagher, R. (eds) *The Human Genome* (Palgrave, Basingstoke, 2001).

Original reference: *Nature* **421**, 444–448 (2003).

Controlling the double helix

Gary Felsenfeld* & Mark Groudine†

*Laboratory of Molecular Biology, National Institute of Diabetes and Digestive and Kidney Diseases, National Institutes of Health, Building 5, Room 212, Bethesda, Maryland 20892-0540, USA (e-mail: gary.felsenfeld@nih.gov)
†Division of Basic Sciences, Fred Hutchinson Cancer Research Center, 1100 Fairview Avenue North, Seattle, Washington 98109, and Department of Radiation Oncology, University of Washington School of Medicine, Seattle, Washington 98195, USA (e-mail: markg@fhcrc.org)

Chromatin is the complex of DNA and proteins in which the genetic material is packaged inside the cells of organisms with nuclei. Chromatin structure is dynamic and exerts profound control over gene expression and other fundamental cellular processes. Changes in its structure can be inherited by the next generation, independent of the DNA sequence itself.

Genes were first shown to be made of DNA only nine years before the structure of DNA was discovered (ref. 1; and see accompanying article by McCarty, page 92). Although revolutionary, the idea that genetic information was protein-free ultimately proved too simple. DNA in organisms with nuclei is in fact coated with at least an equal mass of protein, forming a complex called chromatin, which controls gene activity and the inheritance of traits.

'Higher' organisms, such as yeast and humans, are eukaryotes; that is, they package their DNA inside cells in a separate compartment called the nucleus. In dividing cells, the chromatin complex of DNA and protein can be seen as individual compact chromosomes; in non-dividing cells, chromatin appears to be distributed throughout the nucleus and organized into 'condensed' regions (heterochromatin) and more open 'euchromatin' (see accompanying article by Ball, page 107). In contrast, prokaryotes, such as bacteria, lack nuclei.

The evolution of chromatin

The principal protein components of chromatin are proteins called histones (Fig. 1). Core histones are among the most highly conserved eukaryotic proteins known, suggesting that the fundamental structure of chromatin evolved in a common ancestor of eukaryotes. Moreover, histone equivalents and a simplified chromatin structure have also been found in single-cell organisms from the kingdom Archaeabacteria[2,3].

Because there is more DNA in a eukaryote than in a prokaryote, it was naturally first assumed that the purpose of histones was to compress the DNA to fit within the nucleus. But subsequent research has dramatically revised the view that histones emerged as an afterthought, forced on eukaryotic DNA as a consequence of large genome size and the constraints of the nucleus.

It was known that different genes are active in different tissues, and the distinction of heterochromatin and euchromatin suggested that differences in chromatin structure were associated with differences in gene expression. This led to the early supposition that the histones were also repressor proteins designed to shut off unwanted expression. The available evidence, although rudimentary, does indeed suggest that archaeal histones are not merely packaging factors, but function to regulate gene expression[2–5]. They

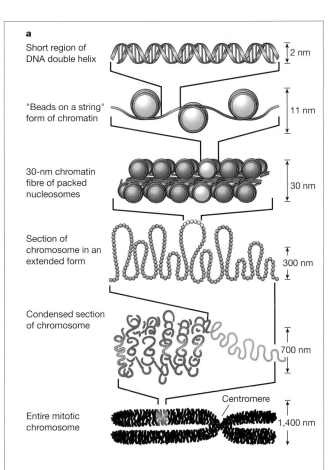

a

Short region of
DNA double helix

2 nm

"Beads on a string"
form of chromatin

11 nm

30-nm chromatin
fibre of packed
nucleosomes

30 nm

Section of
chromosome in an
extended form

300 nm

Condensed section
of chromosome

700 nm

Centromere

Entire mitotic
chromosome

1,400 nm

b

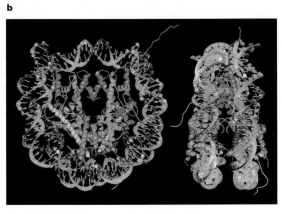

Figure 1 Packaging DNA. **a**, The organization of DNA within the chromatin structure. The lowest level of organization is the nucleosome, in which two superhelical turns of DNA (a total of 165 base pairs) are wound around the outside of a histone octamer. Nucleosomes are connected to one another by short stretches of linker DNA. At the next level of organization the string of nucleosomes is folded into a fibre about 30 nm in diameter, and these fibres are then further folded into higher-order structures. At levels of structure beyond the nucleosome the details of folding are still uncertain. (Redrawn from ref. 41, with permission). **b**, The structure of the nucleosome core particle was uncovered by X-ray diffraction, to a resolution of 2.8Å (ref. 42). It shows the DNA double helix wound around the central histone octamer. Hydrogen bonds and electrostatic interactions with the histones hold the DNA in place.

packaging DNA was an ancillary benefit that was recruited for the more complex nucleosome structure that subsequently evolved in the ancestors of modern eukaryotes, which had expanded genome sizes. Although their compactness might seem to suggest inertness, chromatin structures are in fact a centre for a range of biochemical activities that are vital to the control of gene expression, as well as DNA replication and repair.

Packaging DNA into chromatin

The fundamental subunit of chromatin is the nucleosome, which consists of approximately 165 base pairs (bp) of DNA wrapped in two superhelical turns around an octamer of core histones (two each of histones H2A, H2B, H3 and H4). This results in a five- to tenfold compaction of DNA[6]. The DNA wound around the surface of the histone octamer (Fig. 1) is partially accessible to regulatory proteins, but could become more available if the nucleosome could be moved out of the way, or if the DNA partly unwound from the octamer. The histone 'tails' (the amino-terminal ends of the histone protein chains) are also accessible, and enzymes can chemically modify these tails to promote nucleosome movement and unwinding, with profound local effects on the chromatin complex.

Each nucleosome is connected to its neighbours by a short segment of linker DNA (~10–80 bp in length) and this polynucleosome string is folded into a compact fibre with a diameter of ~30 nm, producing a net compaction of roughly 50-fold. The 30-nm fibre is stabilized by the binding of a fifth histone, H1, to each nucleosome and to its adjacent linker. There is still considerable debate about the finer points of nucleosome packing within the chromatin fibre, and even less is known about the way in which these fibres are further packed within the nucleus to form the highest-order structures.

Chromatin regulates gene expression

Regulatory signals entering the nucleus encounter chromatin, not DNA, and the rate-limiting biochemical response that leads to activation of gene expression in most cases involves alterations in chromatin structure. How are such alterations achieved?

The most compact form of chromatin is inaccessible and therefore provides a poor template for biochemical reactions such as transcription, in which the DNA duplex must serve as a template for RNA polymerase. Nucleosomes associated with active genes were shown to be more accessible to enzymes that attack DNA than those associated with inactive genes[7], which is consistent with the idea that activation of gene expression should involve selective disruption of the folded structure.

Clues as to how chromatin is unpacked came from the discovery that components of chromatin are subject to a wide range of modifications that are correlated with gene activity. Such modifications probably occur at every level of organization, but most attention has focused on the nucleosome itself. There are three general ways in which chromatin structure can be altered. First, nucleosome remodelling can be induced by complexes designed specifically for the task[8]; this typically requires that energy be expended by hydrolysis of ATP. Second, covalent modification of histones can occur within the nucleosome[9]. Third, histone variants may replace one or more of the core histones[10–12].

Some modifications affect nucleosome structure or lability directly, whereas others introduce chemical groups that are recognized by additional regulatory or structural proteins. Still others may be involved in disruption of higher-order structure. In some cases, the packaging of particular genes in chromatin is required for their expression[13]. Thus, chromatin can be involved in both activation and repression of gene expression.

Chromatin remodelling

Transcription factors regulate expression by binding to specific DNA control sequences in the neighbourhood of a gene. Although some DNA sequences are accessible either as an outward-facing segment on the nucleosome surface, or in linkers between nucleosomes, most

may facilitate gene activation, by promoting specific structural interactions between distal sequences, or repression, by occluding binding sites for transcriptional activators.

We suggest that the function of archaeal histones reflects their ancestral function, and therefore that chromatin evolved originally as an important mechanism for regulating gene expression. Its use in

Box 1
Histone modifications

Many amino acids of histones, particularly those in the 'tails', are chemically modified[47]. These include lysine residues that may be acetylated, methylated or coupled to ubiquitin (a large polypeptide chain); arginine residues that may be methylated; and serine residues that are phosphorylated. All modifications can affect one another, and many are positively or negatively correlated with each other. Collectively, they constitute a set of markers of the local state of the genetic material, which has been called the 'histone code'[48].

Histone modification is a dynamic process. Chromatin in the neighbourhood of transcriptionally active genes is enriched in acetylated histones, and the enzymes responsible for both acetylation and deacetylation are often recruited to sites where gene expression is to be activated or repressed, respectively. Within the nucleus, local states of both acetylation and phosphorylation can change rapidly. Methylation at certain histone amino-acid residues may also be important for activation, whereas at other sites it is a signal for inactivation.

Many (perhaps all) of the histone modifications interact with each other in ways that are still not completely understood. For example, in mammals, histone H2B can be modified by ubiquitin at Lys 120 (123 in yeast), and this modification is necessary for methylation at Lys 4 and Lys 79 of histone H3, reactions that are controlled by two different methylating enzymes. Influences between nearby modification sites have also been observed, such that phosphorylation at one site can facilitate acetylation at another, methylation and phosphorylation at adjacent sites may interfere with one another, and methylation and acetylation cannot occur simultaneously on the same lysine residue.

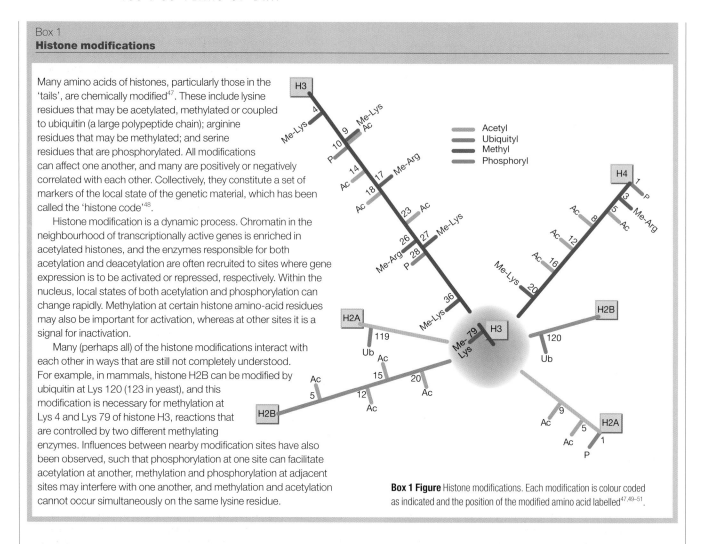

Box 1 Figure Histone modifications. Each modification is colour coded as indicated and the position of the modified amino acid labelled[47,49–51].

---

are buried inside the nucleosome. Regulatory factors must therefore seek out their specific DNA-binding sites and gain access to them. They are aided by chromatin-remodelling complexes that continually shuffle the positions of individual nucleosomes so that sites are randomly exposed for a fraction of time[8,14].

A number of chromatin-remodelling complexes mobilize nucleosomes, causing the histone octamers to move short distances along the DNA[8]. Each complex carries a protein with ATPase activity, which provides the necessary energy. Many of these complexes are members of the so-called SWI/SNF family, which includes SWI/SNF in budding yeast and human, RSC in yeast, and Brahma in *Drosophila*. They have similar helicase-motif subunits, but varying co-factors within the complex. Another SWI/SNF subfamily is based on the helicase-domain protein ISWI, which combines with other proteins to form the complexes NURF, CHRAC and ACF in *Drosophila*, and RSF in humans. A third subfamily is based on the helicase motif protein Mi-2.

Remodelling complexes differ in the mechanisms by which they disrupt nucleosome structure, and they are associated with co-factors that allow them to interact selectively with other regulatory proteins that bind to specific DNA sequences. For example, only certain classes of transcription factors interact with the mammalian SWI/SNF remodelling complex. Thus remodelling complexes can be selective in the genes they modify, and transcription factors recruit these complexes as tools to gain access to chromatin.

Histone modification

Nucleosomes are not passive participants in this recognition process. They can accommodate chemical modifications — either on histone

'tails' that extend from the nucleosome surface, or within the body of the octamer — that serve as signals for the binding of specific proteins. A large number of modifications are already known, such as acetylation of amino acids in the histone tails, and new ones are being identified at a bewildering rate (Box 1). Many modifications are associated with distinct patterns of gene expression, DNA repair or replication, and it is likely that most or all modifications will ultimately be found to have distinct phenotypes.

In addition to histone modifications, nucleosomes can have core histones substituted by a variant, with functional consequences. Histone H2AZ, which is associated with reduced nucleosome stability, replaces H2A non-randomly at specific sites in the genome. Histone H2AX, which is distributed throughout the genome, is a target of phosphorylation accompanying repair of DNA breakage[11], and also seems to be involved in the *V(D)J* recombination events that lead to the assembly of immunoglobulin and T-cell-receptor genes. A histone H3 variant, H3.3, can be incorporated into chromatin in non-dividing cells, and seems to be associated with transcriptionally active genes[10]. Each of these histone substitutions is likely to be targeted by, and associated with, the binding of other proteins involved in gene activation; thus these proteins can be considered central to the formation of localized chromatin structures that are specific for gene activation or accessibility.

Interdependence of histone modifications

An interplay exists between histone modification and chromatin remodelling. For example, expression of a gene may require disruption of nucleosomes positioned at the promoter by a chromatin-remodelling complex before an enzyme required for histone

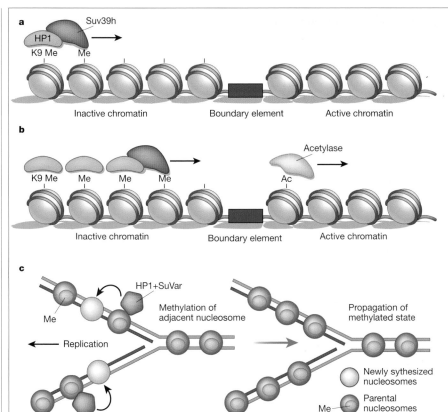

Figure 2 Propagation of inactive ('condensed') and active chromatin states (adapted from ref. 43). **a**, Nucleosomes methylated at H3 Lys 9 are a mark of inactive chromatin and are bound by the heterochromatin protein HP1. HP1 in turn recruits a histone methyltransferase enzyme, Suv39h, that specifically methylates H3 Lys9, allowing methylation and HP1 binding to extend to successive nucleosomes in a self-propagating fashion[43–45]. Some DNA sequence elements (purple rectangle) and their associated proteins may serve as barriers between different chromatin regions, perhaps by blocking the propagation of histone modifications and/or the binding of heterochromatin proteins, thus helping to establish well-defined domains[46]. **b**, A similar propagation mechanism may be constructed for activation by histone acetylation (right). Here, acetylated lysines are recognized by an acetylase enzyme, resulting in acetylation of the adjacent nucleosome. **c**, A proposed model for epigenetic inheritance of methylation. During replication, parental nucleosomes carrying H3 with Lys 9 methylation (blue) are distributed randomly to both sides of the replication fork. Nucleosomes containing newly synthesized histones (pink) are deposited between the old ones, and are methylated by a mechanism similar to that described above. The daughter-cell chromatin then carries the same modification as the parent.

acetylation can be recruited[15]. In contrast, expression of a different gene may require that histone-acetylating enzymes and even RNA polymerase bind to the promoter prior to recruitment of the chromatin-remodelling complex[16]. There is no common series of steps that underlies all or even most processes of gene activation. For any given gene, however, the order of recruitment of chromatin-modifying factors may be crucial for the appropriate timing of expression.

Aside from activating gene expression, histone modifications and chromatin remodelling can also silence genes. Specific histone modifications and chromatin-remodelling complexes, such as the NuRD complex, have been implicated in silencing at some loci[8]. Even SWI/SNF complexes, which are strongly correlated with gene activation, also seem to silence a number of genes.

Specialized chromatin structures

Some regions of the genome are packaged in chromatin with distinct structural features. Three of the most studied such regions are centromeres (important for chromosomal organization during mitosis), telomeres (at the ends of chromosomes) and the inactive X chromosome in mammals. In each case, specific chromosomal structures are defined both by histones modified or substituted in specific patterns, and by the association of additional non-histone proteins or even by regulatory RNA molecules, which increasingly are implicated in chromatin organization[17–19].

Inactive X chromosomes in mammals are enriched for the histone variant macroH2A[20], which is almost three times as large as H2A itself. At vertebrate centromeres, one of the core histones, H3, is replaced by a variant, CENP-A; a similar replacement occurs in centromeres of the fruitfly *Drosophila*, indicating that this is an ancient evolutionary adaptation at centromeres. CENP-A in turn forms a complex with the centromere proteins CENP-B and -C, which mediates the formation of phased arrays of CENP-A-containing nucleosomes. In turn, additional proteins are recruited during cell division to enable the orderly separation of the two chromatids that make up each chromosome. After DNA replication, the sister chromatids are held together initially by a multisubunit complex called cohesin, while a second complex, condensin, helps to compact the chromosomes[21]. These complexes recognize distinct centromere structures, and a specialized nucleosome-remodelling complex associates with cohesin to help it gain access to the chromosomes[22].

In the budding yeast *Saccharomyces cerevisiae*, gene silencing at the ends of chromosomes is mediated by a complex that assembles at telomeres. The complex is stabilized by the binding of the protein RAP1 to the telomere repeat sequences. Additional components, including the silent information regulator (SIR) proteins, then bind inward from the telomere ends, partly through interactions with local nucleosomes[23]. One of the SIR proteins is a histone deacetylase and is thought to repress gene expression at this site. Some components of these unique complexes are evolutionarily conserved, suggesting that these unusual chromatin structures may be found in organisms other than yeast.

The silencing of genes in the vicinity of centromeres in the fission yeast *Schizosaccharomyces pombe* has been shown recently[17–19] to depend on a set of RNA-processing enzymes involved in RNA interference, a process by which double-stranded RNA directs sequence-specific degradation of messenger RNA. One of these enzymes, Dicer, generates RNA fragments about 23 nucleotides long from transcripts of centromeric regions, which then seem in some way to be targeted back to the centromere to initiate the histone-dependent silencing mechanism. Moreover, non-coding RNA transcripts have been identified on the inactive X chromosome and elsewhere in the genome, and may have related roles at those loci[24].

Epigenetic inheritance

An epigenetic trait is one that is transmitted independently of the DNA sequence itself. This can occur at the level of cell division — for example, daughter cells may inherit a pattern of gene expression from parental cells (so-called cellular memory) — or at the generational level, when an offspring inherits a trait from its parents.

The classic example of epigenetic inheritance is the phenomenon of imprinting, in which the expression status of a gene depends upon the parent from which it is derived. In mammals, for example, the *Igf2* gene (encoding insulin-like growth factor-2) is expressed only from the paternal copy of the gene, whereas the *H19* gene is expressed solely from the maternal allele. The mechanism by which this pattern of inheritance is accomplished involves (in part) DNA methylation on the paternal allele. This causes dissociation of a chromatin protein known as CTCF, which normally blocks a downstream enhancer; consequently, the enhancer is then free to activate *Igf2* expression[25,26].

The methylation state of an allele is linked inextricably with patterns of histone modification[27]. Methylated CpG (guanine–cytosine) dinucleotide sites near a gene recruit specific DNA-binding proteins, which in turn recruit histone deacetylases, resulting in loss of histone acetylation and silencing of gene expression. But if histone deacetylation occurs first, it is possible to replace the acetyl group at histone H3 lysine 9 (Lys 9) with one to three methyl groups. It has been shown in turn in the fungus *Neurospora* that the ability to methylate histone Lys 9 is essential for DNA methylation[28], suggesting that local methylation at Lys 9 may provide a signal for methylation of the underlying DNA. Furthermore, in a different reaction pathway, maintenance of histone acetylation at promoters can lead to inhibition of DNA methylation[29].

Epigenetic inheritance involves the maintenance of patterns of histone modification and/or of association of chromosomal proteins correlated with specific expression states. The same mechanisms for propagating permissive or repressive chromatin structure could preserve the pattern of histone modification during replication, when old nucleosomes are distributed randomly on both sides of the fork, with the newly synthesized histones interspersed (Fig. 2).

The maintenance of repressed or activated transcription states represents an efficient mechanism for progressive cellular differentiation[30]. In such a model, fundamental decisions regarding the turning on or off of genes or groups of genes need to be made only once. This principle is perhaps most clearly illustrated by the example of Polycomb-group (PcG)-mediated gene repression in *Drosophila*[31]. At a specific time during development, a complex of proteins, encoded by a collective of PcG genes, binds to sequences within some genes, but only in cells where the genes are silent. At subsequent stages of development, the repressed state is maintained by the PcG complex in the absence of the original negative signals. Activated expression states can be similarly maintained, again in the absence of the original transcriptional activators, by a complex of proteins encoded by genes collectively termed the trithorax group[31]. In both cases, the maintenance of gene-expression patterns is associated with specific histone modification and chromatin-remodelling activities[32–34].

Chromatin and nuclear self-organization

Although bacteria lack a true nucleus, a specific region of the cell, called the nucleoid, contains the chromosome, which in turn is organized into supercoiled domains or loops emanating from central nodes. The organization of the *Escherichia coli* genome into such domains is necessary to allow it to fit within the confines of the cell[2]. Extensions of the chromosome into the cytoplasm correlate with regions that are transcriptionally active. Upon inhibition of transcription, these extensions recede to the nucleoid to give it a more even, spherical shape. The localization of genomic sequences within a bacterial cell is thus determined by their association with the transcriptional/translational apparatus.

The organization of the genome in eukaryotic nuclei, while necessarily more complex than in bacteria, seems to follow the same model as *E. coli*. Individual chromosomes largely occupy distinct 'territories' within the nucleus. Within these territories, actively transcribed genes are on surfaces of channels within subchromosomal domains[35] where soluble transcription factors are presumably more likely to gain access to them.

There is, however, more to the story. The eukaryotic nucleus has distinct subcompartments within which specific nuclear proteins are enriched. For example, the nucleolus, where high-level transcription of ribosomal genes occurs, and splicing-factor compartments accumulate high local concentrations of certain proteins. In some cases there are attachment sites within the nucleus for the proteins. As a rudimentary example, one or more of the proteins associated with yeast telomeres is able to tether the telomeres in clusters to the nuclear periphery[36]. This clustering creates a high local concentration of binding sites for the SIR silencing proteins, which in turn results in a high local concentration of these proteins, and a high occupancy of even relatively weak binding sites. The effect is to increase the extent of telomeric silencing — SIR-dependent gene silencing can be accomplished just by artificially tethering a gene to the nuclear periphery[37].

What organizes the formation of nuclear subdomains? Although there is evidence for a proteinaceous nuclear matrix[38], the example provided by yeast telomeres suggests that the chromatin fibre itself may be the organizer. Many, and probably most, chromatin-binding proteins are in continuous flux between association with chromatin and the nucleoplasm[39,40]. Even such fundamental chromatin proteins as histone H1 have been found to bind for periods of only a few seconds, interspersed with periods of free diffusion. The notable exceptions to this rule are the core histones, the binding of which is much more stable — on the order of minutes for H2A/H2B, and hours for H3/H4. The on–off rates of proteins binding different regions of the genome may depend on the pattern of histone modifications, which in turn determines their relative enrichment in different regions of the nucleus. Thus, the genome as packaged with histones could determine the nature of nuclear subcompartments.

Future challenges

Chromatin proteins and DNA are partners in the control of the activities of the genetic material within cells. The rate-limiting step in activating gene expression typically involves alterations of chromatin structure. The chromosome is an intricately folded nucleoprotein complex with many domains, in which local chromatin structure is devoted to maintaining genes in an active or silenced configuration, to accommodating DNA replication, chromosome pairing and segregation, and to maintaining telomeric integrity. Recent results suggest strongly that in all of these cases the primary indicators of such specialization are carried on the histones. Thus, the regulatory signals that determine local properties, as well as epigenetic transmission of those properties, are likely to be on histones.

The already large catalogue of histone modifications continues to grow rapidly. Although in most cases the loss of the modification (for example, by mutating the responsible enzyme) has a detectable effect on phenotype, the function of many modifications has not yet been determined. While this will be the focus of future research, it presents significant problems because a given modification will occur at many sites in the genome, and mutations could have widespread effects, both direct and indirect. A second significant challenge arises from the potential redundancy of the 'histone code': it is possible that either of two distinct modifications could specify a single structural and functional state, or that the two modifications are always linked to one another. Significant effort will be necessary to determine the complexity of this code, that is, the number of distinct states that can be specified.

The most important immediate problem is to identify the initiating step in establishing a local chromatin state, which may also correspond to an epigenetic state. Silencing at centromeres and perhaps elsewhere seems to be initiated by small RNA transcripts from within the region to be silenced, but formation of other kinds of structures might be triggered directly by a specific histone modification. In the longer term it will be necessary to relate the reactions at individual nucleosomes to higher-order chromatin structures; this will depend in part on the development of higher-resolution methods for determining those structures, and their organization within the nucleus.

At its simplest level, chromatin should be viewed as a single entity, carrying within it the combined genetic and epigenetic codes.

Ultimately our understanding of the dynamic states of chromatin throughout the genome will be integrated with a detailed knowledge of patterns of regulation of all genes.

doi:10.1038/nature01411

1. Avery, O. T., MacLeod, C. M. & McCarty, M. Studies on the chemical nature of the substance inducing transformation of pneumococcal types. Induction of transformation by a desoxyribonucleic acid fraction isolated from Pneumococcus Type III. *J. Exp. Med.* **79,** 137–158 (1944).
2. Sandman, K., Pereira, S. L. & Reeve, J. N. Diversity of prokaryotic chromosomal proteins and the origin of the nucleosome. *Cell Mol. Life Sci.* **54,** 1350–1364 (1998).
3. Ouzounis, C. A. & Kyrpides, N. C. Parallel origins of the nucleosome core and eukaryotic transcription from Archaea. *J. Mol. Evol.* **42,** 234–239 (1996).
4. Bailey, K. A. & Reeve, J. N. DNA repeats and archaeal nucleosome positioning. *Res. Microbiol.* **150,** 701–709 (1999).
5. Dinger, M. E., Baillie, G. J. & Musgrave, D. R. Growth phase-dependent expression and degradation of histones in the thermophilic archaeon *Thermococcus zilligii. Mol. Microbiol.* **36,** 876–885 (2000).
6. Kornberg, R. D. Chromatin structure: a repeating unit of histones and DNA. *Science* **184,** 868–871 (1974).
7. Weintraub, H. & Groudine, M. Chromosomal subunits in active genes have an altered conformation. *Science* **193,** 848–856 (1976).
8. Becker, P. B. & Horz, W. ATP-dependent nucleosome remodeling. *Annu. Rev. Biochem.* **71,** 247–273 (2002).
9. Zhang, Y. & Reinberg, D. Transcription regulation by histone methylation: interplay between different covalent modifications of the core histone tails. *Genes Dev.* **15,** 2343–2360 (2001).
10. Ahmad, K. & Henikoff, S. Histone H3 variants specify modes of chromatin assembly. *Proc. Natl Acad. Sci. USA* 10.1073/pnas.172403699 (2002).
11. Redon, C. *et al.* Histone H2A variants H2AX and H2AZ. *Curr. Opin. Genet. Dev.* **12,** 162–169 (2002).
12. Smith, M. M. Centromeres and variant histones: what, where, when and why? *Curr. Opin. Cell Biol.* **14,** 279–285 (2002).
13. Wolffe, A. P. Nucleosome positioning and modification: chromatin structures that potentiate transcription. *Trends Biochem. Sci.* **19,** 240–244 (1994).
14. Narlikar, G. J., Fan, H. Y. & Kingston, R. E. Cooperation between complexes that regulate chromatin structure and transcription. *Cell* **108,** 475–487 (2002).
15. Cosma, M. P., Tanaka, T. & Nasmyth, K. Ordered recruitment of transcription and chromatin remodeling factors to a cell cycle- and developmentally regulated promoter. *Cell* **97,** 299–311 (1999).
16. Agalioti, T. *et al.* Ordered recruitment of chromatin modifying and general transcription factors to the IFN-β promoter. *Cell* **103,** 667–678 (2000).
17. Hall, I. M. *et al.* Establishment and maintenance of a heterochromatin domain. *Science* **297,** 2232–2237 (2002).
18. Volpe, T. A. *et al.* Regulation of heterochromatic silencing and histone H3 lysine-9 methylation by RNAi. *Science* **297,** 1833–1837 (2002).
19. Allshire, R. RNAi and heterochromatin—a hushed-up affair. *Science* **297,** 1818–1819 (2002).
20. Chadwick, B. P. & Willard, H. F. Cell cycle-dependent localization of macroH2A in chromatin of the inactive X chromosome. *J. Cell Biol.* **157,** 1113–1123 (2002).
21. Nasmyth, K. Segregating sister genomes: the molecular biology of chromosome separation. *Science* **297,** 559–565 (2002).
22. Hakimi, M. A. *et al.* A chromatin remodelling complex that loads cohesin onto human chromosomes. *Nature* **418,** 994–998 (2002).
23. Grunstein, M. Molecular model for telomeric heterochromatin in yeast. *Curr. Opin. Cell Biol.* **9,** 383–387 (1997).
24. Kelley, R. L. & Kuroda, M. I. Noncoding RNA genes in dosage compensation and imprinting. *Cell* **103,** 9–12 (2000).
25. Bell, A. C. & Felsenfeld, G. Methylation of a CTCF-dependent boundary controls imprinted expression of the *Igf2* gene. *Nature* **405,** 482–485 (2000).
26. Hark, A. T. *et al.* CTCF mediates methylation-sensitive enhancer-blocking activity at the *H19/Igf2* locus. *Nature* **405,** 486–489 (2000).
27. Richards, E. J. & Elgin, S. C. Epigenetic codes for heterochromatin formation and silencing: rounding up the usual suspects. *Cell* **108,** 489–500 (2002).
28. Jackson, J. P., Lindroth, A. M., Cao, X. & Jacobsen, S. E. Control of CpNpG DNA methylation by the KRYPTONITE histone H3 methyltransferase. *Nature* **416,** 556–560 (2002).
29. Mutskov, V. J., Farrell, C. M., Wade, P. A., Wolffe, A. P. & Felsenfeld, G. The barrier function of an insulator couples high histone acetylation levels with specific protection of promoter DNA from methylation. *Genes Dev.* **16,** 1540–1554 (2002).
30. Weintraub, H., Flint, S. J., Leffak, I. M., Groudine, M. & Grainger, R. M. The generation and propagation of variegated chromosome structures. *Cold Spring Harb. Symp. Quant. Biol.* **42,** 401–407 (1978).
31. Francis, N. J. & Kingston, R. E. Mechanisms of transcriptional memory. *Nature Rev. Mol. Cell Biol.* **2,** 409–421 (2001).
32. Cavalli, G. & Paro, R. Epigenetic inheritance of active chromatin after removal of the main transactivator. *Science* **286,** 955–958 (1999).
33. Sewalt, R. G. *et al.* Selective interactions between vertebrate polycomb homologs and the SUV39H1 histone lysine methyltransferase suggest that histone H3-K9 methylation contributes to chromosomal targeting of Polycomb group proteins. *Mol. Cell Biol.* **22,** 5539–5553 (2002).
34. Cao, R. *et al.* Role of histone H3 lysine 27 methylation in Polycomb-group silencing. *Science* 298, 1039–1043 (2002).
35. Cremer, T. & Cremer, C. Chromosome territories, nuclear architecture and gene regulation in mammalian cells. *Nature Rev. Genet.* **2,** 292–301 (2001).
36. Laroche, T. *et al.* Mutation of yeast Ku genes disrupts the subnuclear organization of telomeres. *Curr. Biol.* **8,** 653–656 (1998).
37. Andrulis, E. D., Neiman, A. M., Zappulla, D. C. & Sternglanz, R. Perinuclear localization of chromatin facilitates transcriptional silencing. *Nature* **394,** 592–595 (1998).
38. Hart, C. M. & Laemmli, U. K. Facilitation of chromatin dynamics by SARs. *Curr. Opin. Genet. Dev.* **8,** 519–525 (1998).
39. Misteli, T. Protein dynamics: implications for nuclear architecture and gene expression. *Science* **291,** 843–847 (2001).
40. Hager, G. L., Elbi, C. & Becker, M. Protein dynamics in the nuclear compartment. *Curr. Opin. Genet. Dev.* **12,** 137–141 (2002).
41. Alberts, B. *et al. Essential Cell Biology: An Introduction to the Molecular Biology of the Cell* (Garland, New York, 1998).
42. Luger, K., Mader, A. W., Richmond, R. K., Sargent, D. F. & Richmond, T. J. Crystal structure of the nucleosome core particle at 2.8 Å resolution. *Nature* **389,** 251–260 (1997).
43. Bannister, A. J. *et al.* Selective recognition of methylated lysine 9 on histone H3 by the HP1 chromo domain. *Nature* **410,** 120–124 (2001).
44. Lachner, M., O'Carroll, D., Rea, S., Mechtler, K. & Jenuwein, T. Methylation of histone H3 lysine 9 creates a binding site for HP1 proteins. *Nature* **410,** 116–120 (2001).
45. Aagaard, L. *et al.* Functional mammalian homologues of the *Drosophila* PEV-modifier *Su(var)3-9* encode centromere-associated proteins which complex with the heterochromatin component M31. *EMBO J.* **18,** 1923–1938 (1999).
46. West, A. G., Gaszner, M. & Felsenfeld, G. Insulators: many functions, many mechanisms. *Genes Dev.* **16,** 271–288 (2002).
47. Goll, M. G. & Bestor, T. H. Histone modification and replacement in chromatin activation. *Genes Dev.* **16,** 1739–1742 (2002).
48. Strahl, B. D. & Allis, C. D. The language of covalent histone modifications. *Nature* **403,** 41–45 (2000).
49. Wolffe, A. P. & Hayes, J. J. Chromatin disruption and modification. *Nucleic Acids Res.* **27,** 711–720 (1999).
50. van Leeuwen, F., Gafken, P. R. & Gottschling, D. E. Dot1p modulates silencing in yeast by methylation of the nucleosome core. *Cell* **109,** 745–756 (2002).
51. Ng, H. H. *et al.* Lysine methylation within the globular domain of histone H3 by Dot1 is important for telomeric silencing and Sir protein association. *Genes Dev.* **16,** 1518–1527 (2002).

Acknowledgements

We thank the members of our laboratories, L. Hartwell and W. Bickmore for their advice and helpful comments. In particular, we are indebted to M. Bulger for his help in clarifying many of the issues addressed. We also acknowledge the seminal contributions to this field made by our late colleague and friend H. Weintraub, whose ideas continue to serve as a guide to our thinking.

Original reference: *Nature* **421,** 448–453 (2003).

Gene speak

Allele Alternative version of a particular gene. Humans carry two sets of most genes, one inherited from each parent, so a single individual may carry either two of the same, or two different alleles at a given locus, each one inherited separately from one parent.

Amino acid Any of a class of 20 molecules that are combined to form proteins.

Autosome A chromosome not involved in sex determination. The diploid human genome contains 22 pairs of autosomes, and 1 pair of sex chromosomes. Compare with **sex chromosome**.

Base pair (bp) Two nitrogenous bases (adenine and thymine or guanine and cytosine) held together by weak bonds. See **DNA**, **nucleotide**.

Bioinformatics The study of genetic and other biological information using computer and statistical techniques. In genome projects, bioinformatics includes the development of methods to search databases, analyse DNA sequence information, and predict protein sequence and structure from DNA sequence data.

Biotechnology The industry that sprang up in the 1970s from the discovery of bacterial enzymes that cut DNA at specific sites in the genome and rejoin them, enabling human genes, for example, to be 'cloned' in bacteria, and the protein product isolated and purified in large quantities.

Centromere The compact region at the centre of a chromosome.

Chromosome A rod-shaped structure inside the nucleus of a cell which contains a densely packed continuous strand of DNA. Different organisms have different numbers of chromosomes. The diploid human genome consists of 23 pairs of chromosomes, 46 in all: 22 pairs of autosomes and 2 sex chromosomes. See **autosome**, **sex chromosome**.

Clone An exact copy made of biological material, such as a DNA segment (a gene or other region).

Cloning The process of generating multiple, exact copies of a particular piece of DNA to allow it to be sequenced or studied in some other way.

Conserved sequence A sequence of DNA (or an amino acid sequence in a protein) that has remained essentially unchanged throughout evolution, usually because of functional constraints.

DNA (deoxyribonucleic acid) The molecule that encodes genetic information. The four nucleotides in DNA contain the bases: adenine (A), guanine (G), cytosine (C) and thymine (T). Two strands of DNA are held together in the shape of a double helix by bonds between base pairs of nucleotides,

where A pairs with T and G with C. See **nucleotide**, **base pair**.

Diploid A full set of genetic material, consisting of paired chromosomes, one from each parental set. Most animal cells except the gametes have a diploid set of chromosomes. Compare with **haploid**.

Draft sequence DNA sequence in which the order of bases is sequenced at least four to five times (an accuracy of 99.9%), which enables the reassembling of DNA fragments in their original order. Some segments can be missing or in the wrong order or orientation. Compare with **finished sequence**.

Enzyme A protein that specifically catalyses reactions between biological molecules.

Euchromatin The gene-rich regions of a genome. Compare with **heterochromatin**.

Eukaryote An organism whose cells have a complex internal structure, including a nucleus. Animals, plants and fungi are all eukaryotes. Compare with **prokaryotes**.

Excision The process by which enzymes cut out and remove a portion of DNA, for example one that is recognized as containing a mutation.

Exon The protein-coding DNA sequence of a gene. Compare with **intron**.

Finished sequence DNA sequence in which bases are identified to an accuracy of 99.99% and are placed in the right order and orientation along a chromosome with almost no gaps.

FISH (fluorescence *in situ* hybridization) A process that vividly paints chromosomes or portions of chromosomes with fluorescent molecules. This technique is useful for identifying chromosomal abnormalities and gene mapping.

Gamete Mature male or female reproductive cell (sperm or egg) with a **haploid** set of chromosomes.

Gene The fundamental physical and functional unit of heredity. A gene is an ordered sequence of nucleotides located at a particular position on a given chromosome that encodes a specific functional product.

Genetic code The sequence of nucleotides, coded in triplets (codons) along the messenger RNA, that determines the sequence of amino acids in protein synthesis. The DNA sequence of a gene can be used to predict the messenger RNA sequence, and the genetic code can in turn be used to predict the amino acid sequence.

Genome The complete genetic material of an organism; the entire DNA sequence.

Genomics The study of genomes and their sets of genes.

Genotype The set of genes that an individual carries; usually refers to the particular pair of **alleles** that a person has at a given region of the genome. Genotype refers to what is inherited (e.g. an allele for brown eyes), whereas **phenotype** refers to what is expressed (brown eyes in this case).

Haploid A single set of chromosomes (half the full set of genetic material) present in the egg

and sperm cells of animals and in the egg and pollen cells of plants. Compare with **diploid**.

Haplotype A particular combination of **alleles** or sequence variations that are closely linked – that is, are likely to be inherited together – on the same chromosome.

Heterochromatin Compact, gene-poor regions of a genome, with abundant simple sequence repeats. Compare with **euchromatin**.

Intron The DNA sequence interrupting the protein-coding sequence of a gene; this sequence is transcribed into RNA but is cut out before the RNA is transcribed. Compare with **exon**.

Karyotype An arrangement of an individual's chromosomes in a standard format showing the number, size and shape of each chromosome type; used in low-resolution physical mapping to correlate gross chromosomal abnormalities with the characteristics of specific diseases.

Kilobase (kb) A unit of length of DNA fragments equal to 1000 nucleotides.

Library A unordered collection of clones whose relationship to each other can be established by physical mapping.

Linkage The proximity of two or more markers (e.g. genes) on a chromosome.

Marker An identifiable physical location or landmark on a chromosome (e.g. a restriction enzyme cleavage site) whose inheritance through generations can be monitored.

Megabase (Mb) Equal to 1 million nucleotides.

Meiosis The process of two consecutive cell divisions in the diploid progenitors of sex cells, which results in four progeny cells, each with a **haploid** set of chromosomes.

mRNA (messenger RNA) The form of RNA that serves as a template for protein synthesis.

Mitosis The process of nuclear division in cells that produces two daughter cells genetically identical to each other and to the parent cell.

Mutation An alteration in the sequence of DNA, for example the substitution of one nucleotide base for another.

Nucleotide Subunit of DNA or RNA comprising a nitrogenous base (adenine, guanine, thymine or cytosine in DNA; adenine, guanine, uracil or cytosine in RNA), a phosphate molecule and a sugar molecule (deoxyribose in DNA and ribose in RNA). Nucleotides are linked to form the strands of a DNA or RNA molecule. Short stretches of nucleotides are called oligonucleotides. See **base pair**.

PCR The polymerase chain reaction: a technique that produces millions of copies of a short stretch of DNA in a matter of hours, providing enough material for the detection, for example, of a single mutation in a biological specimen such as a single cell taken from a human embryo.

Peptide A sequence of amino acids that is shorter than a protein.

Phenotype The observable properties and physical characteristics of an organism.

Polymorphism A difference in DNA sequence between individuals. To be called a polymor-